ASTRONOMICAL HAZARDS FOR LIFE ON EARTH
IAU SYMPOSIUM 374

COVER ILLUSTRATION:

Asteroid impact: Art courtesy David A. Hardy, https://url.uk.m.mimecast protect.com/s/TbHJCBLoos71AnG5izfRU2I429?domain=astroart.org
Sun & CME impacting Earth: many sources in internet Supernova explosion reaching Earth: NASA

IAU SYMPOSIUM PROCEEDINGS SERIES

Chief Editor
JOSÉ MIGUEL RODRIGUEZ ESPINOSA, General Secretariat
Instituto de Astrofisica de Andalucía
Glorieta de la Astronomia s/n
18008 Granada
Spain
IAU-general.secretary@iap.fr

Editor
DIANA WORRALL, Assistant General Secretary
HH Wills Physics Laboratory
University of Bristol
Tyndall Avenue
Bristol
BS8 1TL
UK
IAU-assistant.general.secretary@iap.fr

INTERNATIONAL ASTRONOMICAL UNION

UNION ASTRONOMIQUE INTERNATIONALE

International Astronomical Union

ASTRONOMICAL HAZARDS FOR LIFE ON EARTH

PROCEEDINGS OF THE 374th SYMPOSIUM OF THE INTERNATIONAL ASTRONOMICAL UNION BUSAN, REP OF KOREA 9–11 AUGUST 2022

Edited by

GONZALO TANCREDI

Div. F Planetary Systems and Astrobiology, UdelaR, Uruguay

CAMBRIDGE
UNIVERSITY PRESS

CAMBRIDGE UNIVERSITY PRESS
University Printing House, Cambridge CB2 8BS, United Kingdom
1 Liberty Plaza, Floor 20, New York, NY 10006, USA
10 Stamford Road, Oakleigh, Melbourne 3166, Australia

First published 2025

Printed in Great Britain by Henry Ling Limited, The Dorset Press, Dorchester, DT1 1HD

Typeset in System LaTeX 2ε

*A catalogue record for this book is available from the British Library Library of Congress
Cataloguing in Publication data*

This journal issue has been printed on FSC^{TM}-certified paper and cover board. FSC is an
independent, non-governmental, not-for-profit organization established to promote the
responsible management of the world's forests. Please see www.fsc.org for information.

ISBN 9781009353083 hardback
ISSN 1743-9213

For EU product safety concerns, contact us at Calle de José Abascal, 56, 1°, 28003 Madrid,
Spain, or email eugpsr@cambridge.org

Table of Contents

Preface

The end of humanity has been a topic of great concern across ages and civilizations. This is reflected in the wealth of references throughout many cultures and religions. Over the last several decades, studies have allowed us to better understand the most likely threats to life on Earth, both in the past and the future. This symposium focused on a comparative analysis of natural threats, caused by astronomical phenomena, which could lead to a new extinction, and not the anthropic causes. Current and future mitigation strategies were also be discussed.

During the Symposium, we covered several potential hazards caused by astronomical phenomena. Thus, this is multi- and cross-disciplinary topic, encompassing almost all IAU Divisions.

Although the problem of astronomical risks for life on Earth has been a matter of concern to some IAU bodies, such as the WG Near Earth Objects, or Commission E3 Solar Impact Throughout the Heliosphere and Inter-Division E-F-G Commission Impact of Magnetic Activity on Solar and Stellar Environments, is the first time that scientists from various Divisions meet to analyze the problem as a whole.

The symposium was organized in different sessions, devoted to the following topics:

- Terrestrial hazards: Earth magnetic field
- Planetary hazards: asteroid and comet impacts, rogue planets
- Solar hazards: solar activity, Sun evolution
- Galactic hazards: nearby stars, heliosphere, supernovae, GRBs, black holes
- Universal hazards: the fate of the Universe

An important part of the presentations were associated with the risk of impact of asteroids and comets against the Earth, and its consequences; nevertheless, there were also presentations on solar hazards, the supernovae threat, stellar close encounters, and the fate of the Universe.

There were two Round Tables with the participation of several of the invited speakers. The first one was on "Comparative analysis of the astronomical hazards". The different threats were analyzed in comparative terms, assessing the relevance of each one of them. The participants also discussed the mitigation actions that humanity has been developing to face these problems.

The second Round Table was titled: "Dealing with the hazards: the role of scientists, public, media and decision makers". The initiative promoted by a group of colleagues of various nationalities and expertise to propose the declaration of 2029 as the International Year of Planetary Defense, by the UN, was publicly presented for the first time. The difficult dialogue between the scientific community, the public, the press and decision makers in the face of a specific threat was analyzed.

Unfortunately, due to the health situation, in which travel was still highly restricted, and the time difference with the Western Hemisphere, the number of in-room and remote participants was much lower than expected.

We hope that these topics can continue to be analyzed in inter-Divisions discussions within the IAU.

SOC Members

- Gonzalo Tancredi: Div. F Planetary Systems and Astrobiology, UdelaR, Uruguay (Chair)
- Daniel Hestroffer: Div. A Fundamental Astronomy, Obs. Paris, France
- Pete Riley: Div. E Sun and Heliosphere Predictive Science, San Diego, California, USA
- Isabelle Grenier: Div. D High Energy Phenomena and Fundamental Physics, CEA Saclay, AIM, Service d'Astrophysique, France
- Monica Rubio: Div. H - Depto. Astronomía, Universidad de Chile, Chile
- Gustavo Bruzual - Div. J - Instituto de Radioastronomia y Astrofísica, UNAM, México
- Romana Kofler: Programme Officer Committee, policy and legal affairs section United Nations Office for Outer Space Affairs (UNOOSA)
- Doris Daou: Planetary Defense Coordination Office - US
- Detlef Koschny: ESA Planetary Defence Office, Netherlands
- Boris Shustov: Russian Academy of Sciences Institute of Astronomy, Russia
- Makoto Yoshikawa: JAXA, Japan
- Natchimuthuk Gopalswamy: Director International Space Weather Initiative (ISWI) Solar Physics Laboratory, NASA/GSFC, US
- Annapurni Subramaniam Director of Indian Institute of Astrophysics, India
- Paul Gabor: Vatican Observatory, Vatican City State
- Francisco O'Reilly: Philosopher, Universidad de Montevideo, Uruguay

List of Participants

List of participants, including the country.

This is the list of contributors (oral and poster), because since the Symposium was held during the GA, I do not have the information of the specific participants for the Symposium.

Name	Country	Name	Country
Ammar Abdulla	Arab Emirates	Thomas Statler	USA
Brent Barbee	USA	Vasiliki Petropoulou	Italy
Brian Thomas	USA	Xuguang Leng	USA
Camilo Delgado-Correal	Colombia	Coryn Bailer-Jones	Germany
Diederik Kruijssen	Germany	Christopher Impey	USA
Elisabetta Dotto	Italy	Milan Cirkovic	Serbia
Gijs Verdoes Kleijn	Netherlands	Anne-Charlotte Perlbarg	France
Gonzalo Tancredi	Uruguay	Boris Shustov	Russia
Gulchehra Kokhirova	Russia	Eduard Kuznetsov	Russia
Heidi Korhonen	Chile	Fabrizio Bernardi	Italy
Ihor Kyrylenko	Russia	Joseph Masiero	USA
James Bauer	USA	Roman Zolotarev	Russia
Jeffrey Love	USA	Sun Mie Park	Republic of Korea
Makoto Yoshikawa	Japan	Supachai Awiphan	Thailand
Patrick Michel	France	Yudish Ramanjooloo	USA
Richard Wainscoat	USA	Birgit Loibnegger	Austria
Rosita Kokotanekova	Bulgaria	Sergei Ipatov	Russia
Simone Ieva	Italy	Remziye Canbay	Turkey
Suresh Bhattarai	Nepal	Gulchehra Kokhirova	Russia
Svitlana Kolomiyets	Ukraine	Eva Villaver	Spain
Teymoor Saifollahi	Iran		

Astronomical Hazards for Life on Earth
Proceedings IAU Symposium No. 374, 2025
G. Tancredi, ed.
doi:10.1017/S1743921324000784

What catastrophes can affect us at different geographical and temporal scales?

Gonzalo Tancredi ⓘ

Departamento de Astronomía, Facultad de Ciencias, Udelar, Uruguay.
email: `gonzalo@fisica.edu.uy`

Abstract. Life on Earth can be (has been) affected by various phenomena of non-biological or extraterrestrial origin, such as worldwide volcanic eruptions, the impacts of asteroids and comets, solar storms, the evolution of the Sun, supernova explosion, cosmic ray showers, ...

These phenomena and the associated catastrophes can be grouped into different categories depending on its origin: terrestrial, solar system, galactic, and extragalactic, and the final destiny of the Universe. We will shortly described many of the identified risks and compare them by the degree of affectation for Life and Humanity.

The time scales and the area on Earth affected by each phenomenon vary considerably among them. We list the phenomena that can affect a region the size of a country, a continent, or a global catastrophe. However, we note that, given humanity's degree of global economic and social interdependence, a local-scale phenomenon can even have global consequences.

The risks can be further classified in random and deterministic. Random threats are those associated with an event that has a certain probability of occurrence on a time scale, but we do not the exact date in the future, i.e.: an asteroid impact or a supernova explosion. Deterministic threats are those that will surely occur in a range of time in the future, i.e.: the evolution of the Sun into a red giant.

This comparative study will analyze what the "certainties" are (in statistical terms) about the different phenomena of extraterrestrial origin that will affect life on Earth on different geographical and temporal scales.

Keywords. catastrophes, impacts, solar storms, supernovae

---

1. Introduction

Before we get into the analysis of the different catastrophes that can affect life on planet Earth, we will make some terminological definitions.

A hazard or threat is the potential to cause a harm, while a risk is the likelihood of a certain harm taking place. A car is a hazard when crossing a road. But the risk of an accident with a car depends on the amount of traffic, it is larger in a highway than in a lonely road.

The list of catastrophic threats that can affect life on Earth is very long. Hereby we are not considering the threats from biological origin (e.g. pandemia), not even the anthropogenic ones (e.g. wars). Among the non-biological threats, we group them according to their origin: terrestrial, solar system, galactic, extragalactic and universal.

Hereby is a list, as comprehensive as possible, of the potential threats in a large geographical scale, even planetary:

- Terrestrial threats: coming from the components of the Earth System.
  Mega-Tsunamis
  Mega-Earthquakes

> Worldwide volcanic eruptions and basalt floods
> Universal flooding
> Non-anthropogenic Greenhouse effect
> Variation of the Earth's magnetic field
> Tides and the rotation of the Earth-Moon System

- Solar System threats: coming from the members of the Solar System.
  > **Comet and asteroid impacts**
  > Giant comet in the inner Solar System
  > Comet shower
  > Stellar companion of the Sun (Nemesis)
  > **Solar activity**
  > **Evolution of the Sun**

- Galactic threats: coming from tthe members of our galaxy, the Milky Way.
  > Heliosphere and the Local Bubble
  > **Passage of nearby stars**
  > Passage of the Sun through the galactic plane
  > Passage of the Sun through the spiral arms
  > **Nearby Supernovae**
  > **Gamma-ray Bursts and Hypernovae**
  > Drifting Black Holes
  > Black Hole at center of the Galaxy

- Extragalactic threats: coming from objects outside our galaxy.
  > **Collision of the Milky Way and Andromeda Galaxy**
  > Approach to the Virgo Supercluster

- Universal threats: with this we refer to the final destiny of the Universe.
  > **Halt of star formation and isolation**
  > The Geometry of the Universe
  > Big crunch
  > **Big Rip**

A detailed description of the above listed threats is out of this scope of the comparative analysis considered in this article. Many of those threats have been described in several books, and the interested reader can refer to them, like:

- *Global Catastrophic Risks* - Ed. N. Bostrom & Ćirković 2008
- *How it ends* - C. Impey 2010
- *Death from Skies!: These Are the Ways the World Will End...* - P. Plait 2008
- *Cataclysms: A New Geology for the Twenty-First Century* - M. Rampino 2017
- *Global Catastrophes: A Very Short Introduction*, B. McGuire 2014
- *Cosmic Catastrophes - Supernovae, Gamma-ray Bursts, and Adventures in Hyperspace* - J. Craig Wheeler 2000
- *The Ends of the World: Volcanic Apocalypses, Lethal Oceans, and Our Quest to Understand Earth's Past Mass Extinctions* - P. Brannen 2018
- *The End - What Science and Religion Tell Us about the Apocalypse* - P. Torres 2016
- *Perilous Planet Earth: Catastrophes and Catastrophism through the Ages* - T. Palmer 2003

It is also out of the scope of this article a review of the scientific merits of the above books. But, after analyzing them as well as many other articles, we found that there is no literature where the hazards have been analyzed in comparison, as well as a comparative analysis of the corresponding risks. The objetive of this article is a first approach to this issue.

The threats listed above can affect us on a very wide range of spatial and temporal scales. In addition to a classification according to the origin of the phenomena causing the risk, they can also be classified according to its probabilistic nature into two main types:

- random threats: those associated with an event that has a certain probability of occurrence on a time scale.
- deterministic hazards: are those that will occur over a range of time in the future.

Random threats include phenomena such as the impact of an asteroid or comet, the arrival of a coronal mass ejection from the Sun, the close passage of a star, or the explosion of a supernova. Within the deterministic threats we have the evolution of the Sun as a red giant, the collision of the Milky Way and Andromeda, or the final destiny of the Universe.

Deterministic threats will occur for sure; we generally have an estimate of the time in the future when the event will occur.

For random hazards, on the other hand, we know that the event will occur with a certain frequency, but we cannot say exactly when. These types of hazards are known phenomena with a certain degree of predictability. For example, if we know the orbits of the asteroids that cross the Earth's orbit, we will know if it has a chance of impacting the Earth in the next few decades. Or, if we determine the position and velocity of the stars approaching the Sun, we will know if any of them may come within a dangerous distance in the next few thousand years. Or, if we monitor nearby massive stars, we will know if one has a chance of exploding as a supernova when it is close to us. So, we generally have a period in the future time when we have (near-) certainty that an event may not occur (window of predictability): no asteroid will hit in the next few decades, no star will approach the Sun within 1 pc in the next few thousand years, no nearby supernova will explode in the next few million years. But after that period has passed, we only have a probabilistic estimate. Nevertheless, we can define a period of time when the probability of occurrence of the event is so high that, for practical purposes, it can be considered that an event should occur.

Let us consider a phenomenon with an average frequency of occurrence of an event every T number of years; T is the average time span separating two consecutive events. The occurrence of this type of random event can be considered as a Poisson process. A Poisson process has the property that each event is stochastically independent of all other events in the process, so that the occurrence of an event is completely random. For example, phenomena that can be considered as a Poisson process are: the arrival of calls to a telephone call center, the passage of cars through a tollbooth, the emission of particles in a radioactive decay, the occurrence of earthquakes, etc.

In a Poisson process, in an infinitesimal time interval dt only one event can occur, and this happens with a probability equal to λdt, where $\lambda = 1/T$. This probability is independent of events outside the interval. The probability of an event at any particular instant is 0. The probability that the number of events $N(t)$, in a finite interval of length t, be equal to n is given by the Poisson distribution:

$$P\left\{N(t) = n\right\} = \frac{(\lambda t)^n}{n!} e^{-\lambda t} \tag{1}$$

(the left-hand term of the equation is read as: the probability that $N(t)$ equals n is equal to ...).

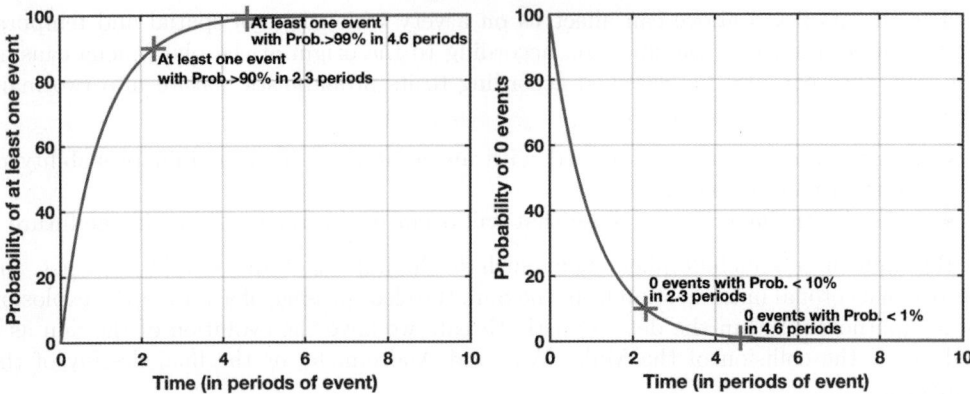

Figure 1. *a* Probability of occurrence of at least one event as a function of time, expressed in periods of event. *b* Probability of occurrence of zero events as a function of time.

The probability of occurrence of at least one event in a period of time t is equal to the summation of the above expression for values of $n = 1, ..., \infty$; that is:

$$P\left\{N(t) \geq 1\right\} = \sum_{n=1}^{\infty} P\left\{N(t) = n\right\} = 1 - e^{-\lambda t}. \tag{2}$$

While the probability of no event occurring at time t is equal to 1 minus the above probability; that is:

$$P\left\{N(t) = 0\right\} = e^{-\lambda t}. \tag{3}$$

In Fig. 1 *a* and *b* the probabilities given by eqs. 1 and 2 are plotted.

We will consider as "quasi-certainty of occurrence" of the event, when there is a very high probability of occurrence in a certain time scale. For example, if we assume the value of 90% probability, we will say that in a time equivalent to 2.3 typical periods of the event, there is a quasi-certainty that the phenomenon will occur, with a probability higher than 90%. If we want to increase the range of certainty, and consider a 99% probability for the event to occur, we require a time equivalent to 4.6 typical periods.

2. Comparative analysis

With these considerations in mind, we will proceed to a comparative analysis of the most relevant phenomena listed above on different time scales. The threats considered in this analysis are those listed in bold in the above list.

2.1. Certainties on the scale of centuries or a few thousand years

- **A solar storm and coronal mass ejection (CME) will occur, affecting the vast majority of artificial satellites and impacting much of the electronic equipment on the Earth's surface.** The generation of CMEs follows the 11-year pattern of solar activity. It is estimated that an event similar to the Carrington event of 1859 may occur every few centuries; therefore, we can state with great certainty that an event of this type will occur in the next millennium. The arrival of a coronal mass ejection on Earth would not cause a mass extinction phenomenon, but the consequences for the technological equipments developed by mankind would be extremely serious. With the continuous monitoring of the solar activity and early detection of CMEs leaving the Sun, the arrival of a CME can be predicted with 2–3 days in advance. However, at present we have no way to mitigate the effects of the arrival of an intense CME, beyond a programmed technological blackout to reduce the propagation of induced currents.

- **An object of a few tens of meters in size will impact the Earth with catastrophic consequences for an area equivalent to a large city or small province** (a few thousand km^2). We are refering to an event similar to the Tunguska event in 1908. The frequency of a Tunguska-type event is one event every few hundred years (200-1000 yr). Let's consider that on Earth there are $\sim$1000 population centers with more than half a million inhabitants, the city has to be within the $\sim$10,000 km^2 area that is affected by a Tunguska-type event; then, the total area on Earth exposed to an impact affecting a densely populated area is $10^3 \times 10^4 = 10^7$ km^2. This represents 2% of the total surface area of the Earth. Therefore, we should expect a Tunguska-type event to affect a densely populated area, typically, every few tens of thousands of years (with 90% of confidence, the period is $\sim 1/0.02 \times 500 \times 2.3 = 6 \times 10^4$ yr). Therefore, with high certainty we can state that some city on Earth could be destroyed in the next 60,000 years as a consequence of an impact. If the trend of population growth continues to increase, and there is an increase in the number of population centers, this time interval would be reduced. With the asteroid and comet deflection technologies that have been developed in recent years, we are in a position to deflect an asteroid of a few meters in diameter. The challenge is to discover these small objects of very faint brightness early enough to have enough time to launch a deflection mission.

2.2. Certainties on a scale of several hundred thousand to millions of years (Mega-years= 10^6 yr).

- **Reversal of the Earth's magnetic field will occur**, passing through a brief period (possibly thousands of years) in which the magnetic field strength will be very low. The average interval between magnetic field reversals over the past few million years has been 250,000 years; and the last reversal occurred 780,000 years ago. While there is not yet an accepted model for the causes of the reversals, statistically speaking, we should expect a new reversal to occur in the next few hundred of thousands years. The "magnetic shield", which protects us from the flow of energetic particles from the Sun, will disappear or will be significantly reduced. The Earth will be more exposed to the solar wind and coronal mass ejections. The flux of ionizing radiation, from both solar and galactic energetic particles, will increase in the atmosphere and at the surface. As a consequence of this increase, chemical reactions will occur in the atmosphere, such as the destruction of ozone and an increase in UV radiation received at surface level. There will be increased damage to electronic circuits at surface level and in low orbits. Biological tissues will be exposed to ionizing radiations producing an increase in the rate of mutations. We still need to better understand the generation and evolution of the Earth's magnetic flied in order to make reliable predictions on when the new reversal could occur.

- **There will be an asteroid or comet of more than 1 km in size impacting the Earth, that could produce a catastrophe with consequences on a global scale.** Such an event will almost certainly occur within the next ten million years. An impact that would produce a mass extinction of life on Earth would be caused by an asteroid larger than 5 km in diameter, and that could occur in less than 100 million years. At present we have already catalogued over 95% of the $\sim$ 900 existing asteroids in the size range over 1 km, and none of them will impact the Earth in the next decades. Prediction on longer scales are difficult due to the observational errors and the chaotic nature of their motions. There are less than 50 asteroids to be discovered in this size range, for which we can not make any prediction. With a continuos monitoring and survey of the population of Near-Earth Asteroids, we can have many decades of alerts before an object potentially collide with the Earth. Several deflection mechanisms could be used with a long period of alert.

- **A star will approach the Sun at a distance that will produce a major disturbance in the cloud of comets surrounding the Solar System (the Oort cloud).** The flux of new comets will increase for a few million years, increasing the probability of impacts with the Earth. Such a passage should occur within the next ten million years. With the new astrometric data from the ESA-Gaia space mission, we know the motion of all the nearby stars, and we can make predictions of future passages.

- **A nearby supernova star will explode, throwing an intense flux of high-energy radiation and energetic particles into the Solar System.** For a supernova explosion to have catastrophic consequences for life on Earth, its distance should be less than ~10 parsecs, at the time of the explosion. The atmosphere, and in particular the ozone layer, will be strongly perturbed. A supernova explosion at a distance less than 10 parsecs from the Sun should occur within the next hundred million years. Gaia also helped to identify the position and motion of supernova candidates. Nevertheless, the stellar evolution models are not precise enough to forecast when a giant star could become a supernova in a short-time scale.

2.3. Certainties on the scale of billions of years (Giga-years= $10^9 yr$)

- **A Gamma Ray Bursts (GRBs) emitting a narrow high energy flux will hit the Earth.** GRBs are emissions of electromagnetic radiation in the form of gamma rays of very high energy and very short duration (typically a few seconds). They are associated with the death of massive stars or the merge of a binary neutron stars. The gamma-ray burst that could affect us, should occur at a distance of less than 3 kiloparsecs from the Sun, with an emission cone pointing in the direction to the Solar System. A very high energy radiation flux will be received with consequences at the level of the Earth's atmosphere and surface. Such an event should occur within the next Giga-years.

- **The Milky Way Galaxy will approach the Andromeda Galaxy and they could possibly collide.** The event could increase the chances of collision or close passage among the stars in both system, and in particular a close approach with the Solar System. The galactic approach will occur in ~4 Ga; but the geometry of the encounter is uncertain, as well as the consequences for our Solar System. Again the astrometric information from Gaia will be useful to reveal the circumstances of the close approach.

- **The Sun will evolve into a red giant star.** In 5 Giga-yr the average temperatures on Earth will be higher than the maximum temperatures at present. The expansion of the Sun will continue until it engulfs the Earth in 7.6 Giga-yr. All forms of life will be unviable on our planet.

2.4. Certainties on very large time scales (greater than Tera-years= $10^{12} yr$)

- **Star formation will come to an end within 1-100 Tera-years**, since there will be no interstellar gas in sufficient densities to form new stars by cloud collapse.

Stellar corpses, such as white dwarfs, will cool down and become black dwarfs. There will be no energy sources to support any form of life.

- **The destiny of the Universe.** Measurements of cosmological parameters point to a Universe that will continue to expand at an accelerated rate, decreasing its temperature. It will become a dark and cold Universe. According to this model, at a certain finite time in the future, there will be a final singularity, in which the distance scale of the universe will diverge to infinite values, producing the tearing apart of matter (the Big Rip). However, the measurements of the cosmological parameters still have such a large uncertainty that it is not possible to be certain about the final fate of the Universe and the corresponding time scales.

3. Final considerations

Humanity is the first specie on Earth capable of identifying the variety of terrestrial and astronomical phenomena that can produce large-scale catastrophes. Most of the threats considered above have only been recognized in recent decades. Although the fear of a catastrophic event that could end life has been a common concern of many cultures, and they have been expressed in numerous apocalyptic stories, until the advancement of our scientific knowledge, it has not been possible to adequately analyze the relevance of these ideas.

In addition to identify the threats, we are now in a position to estimate the risks associated with them, and to make a comparison among them. Furthermore, we have started to make plans of possible mechanism to mitigate the effects of catastrophes, and even to test the technologies that could be used in case of an imminent event could happen, like the NASA-DART experiment. The collaboration among word-wild partners like in the International Asteroid Warning Network (IAWN), the Space Mission Planning Advisory Group (SMPAG), and the International Space Weather Initiative (ISWI) are fundamental to tackle the problem together.

In less than a century, a very short time scale compared to most of the threats analyzed in this article, we have advanced a lot on protecting life on Earth. In this issue, we are leaving a more secure planet for our descendants and other species.

However, we know that the hazards discussed in this article are not the only ones that threaten life on our planet. As mentioned at the beginning, in this article we concentrate in the non-biological threats for life. We cannot forget that the largest threat at present for the entire planet is our humankind. It is our responsibility not to be the cause of a new mass extinction.

References

Brannen P., 2018, *The Ends of the World: Volcanic Apocalypses, Lethal Oceans, and Our Quest to Understand Earth's Past Mass Extinctions*, Ecco.

Bostrom N. & Ćirković M., 2008, *Global Catastrophic Risks*, Oxford University Press.

Impey C., 2010, *How it ends*, W.W. Norton & Company, New York-London.

McGuire B., 2014, *Global Catastrophes: A Very Short Introduction*, Oxford University Press.

Palmer T., 2003, *Perilous Planet Earth: Catastrophes and Catastrophism through the Ages*, Cambridge University Press.

Plait P., 2008, *Death from Skies!: These Are the Ways the World Will End...*, Viking Penguin.

Rampino M., 2017, *Cataclysms: A New Geology for the Twenty-First Century* - Columbia University Press.

Torres P., 2016, *The End - What Science and Religion Tell Us about the Apocalypse*, Pichstone Publishing, Durham, North Carolina.

Wheeler J. Craig, 2000, *Cosmic Catastrophes - Supernovae, Gamma-ray Bursts, and Adventures in Hyperspace*, Cambridge University Press.

Astronomical Hazards for Life on Earth
Proceedings IAU Symposium No. 374, 2025
G. Tancredi, ed.
doi:10.1017/S1743921324000723

The Pan-STARRS Search for Near-Earth Objects

Richard Wainscoat[ID], Mark Huber, Robert Weryk, Yudish Ramanjooloo, John Fairlamb, Kenneth Chambers, Eugene Magnier, Thomas de Boer, Chien-Cheng Lin and Hua Gao[ID]

Institute for Astronomy, University of Hawaii, Honolulu, Hawaii, USA
email: rjw@hawaii.edu

Abstract. The Pan-STARRS telescopes, located on Haleakala, Maui, Hawaii, are conducting a long-term search of the night sky for Near-Earth Objects. The Pan-STARRS survey is now one of the leading NEO surveys, accounting for more than 40% of all new NEO discoveries, and over 50% of discoveries of larger NEOs. Pan-STARRS is also a prolific comet discovery telescope, and discovered the first interstellar object, 'Oumuamua.

Keywords. Near-Earth Object, asteroid, comet, Earth impact hazard, survey

1. Introduction

Near-Earth Objects (NEOs) are defined as objects with perihelion less than 1.3 au. They can be asteroids or comets. The US congress instructed NASA to discover at least 90% of all NEOs with diameter > 1 km, a goal that is now believed to have been achieved Mainzer *et al.* (2011). This has since been extended to a goal of finding at least 90% of all NEOs with diameter greater than 140 m. In 2022, approximately 10,000 NEOs with diameter > 140 m had been discovered. This corresponds to approximately 40% of the expected number ($\sim 25,000$) of these objects Harris and d'aBramo (2015). Contemporary surveys, including Pan-STARRS, are discovering these > 140 m sized asteroids at a rate of approximately 500 per year, or 2% of the population.

The Pan-STARRS telescopes are 1.8-meter wide-field telescopes located near the summit of Haleakala, on the island of Maui in Hawaii. Haleakala is the best site in the world for solar astronomy, and is also an excellent site for night time astronomy — arguably the second best site in the United States. For night time astronomy, Haleakala is inferior to Mauna Kea on the adjacent island of Hawaii. Haleakala is lower than Mauna Kea, which means that low-level moisture can more easily go over the summit. Haleakala has a large crater to the northeast of the summit that can produce turbulence with the prevailing northeast winds and make the seeing on Haleakala inferior to Mauna Kea. Maui is a smaller island with people living closer to the observatory and unfortunately has a weaker lighting ordinance than the island of Hawaii, resulting in a slightly brighter night sky.

Each telescope has a large (0.9 m) secondary mirror and a very large camera mounted at the Cassegrain focus. The Pan-STARRS1 (PS1) camera has almost 1.4 billion pixels, and the Pan-STARRS2 (PS2) camera has almost 1.5 billion pixels. These cameras are presently the largest digital cameras in the world, but will be surpassed by the camers on the Rubin Observatory when it becomes operational.

The PS1 camera has 60 CCDs, and the PS2 camera has 64 CCDs. These are arranged in an 8×8 array, with the corners empty in the PS1 camera. The CCDs are Orthogonal Transfer Arrays (OTAs), each with an 8×8 grid of 600×600 active pixels and an inactive grid between the active cells. The intent of this design was to move charge around in the device to compensate for telescope shake to improve image quality. This functionality is not normally used, because it causes cosmetic defects in the devices to grow as the charge is moved around. The complexity of the fabrication of the OTAs resulted in devices that had more cosmetic defects than normal CCDs, and some areas of poor charge transfer efficiency that are unusable. The effective fill factor of each camera is a relatively poor 70%. The PS1 camera suffers from an image persistence problem when pixels have been saturated. The CCDs in the PS2 camera do not suffer from this problem. The OTA grid pattern is not ideal for discovery of moving objects because objects can move from the active cells into the inactive grid surrounding the cells.

Each camera has a field-of-view that is approximately 3 degrees in diameter, and 7 deg^2 in area. Pixels in each camera are approximately 0.25 arc seconds. Typical seeing on Haleakala is approximately 1.0 arc seconds. Good seeing can be as good as 0.75 arc seconds, and usually coincides with light winds, of southerly winds. Both telescopes are operated remotely from the Institute for Astronomy office in Pukalani, Maui, which is approximately a 1-hour drive from the telescopes. Telescope operators can drive to the summit if there is an emergency need, but in practice, this is seldom necessary.

PS1 began operation in 2010 and spent its first 4 years doing a multi-purpose multi-color survey of the sky north of $-30°$ declination. The survey was completed in March 2014, and PS1 has been conducting a survey for Near-Earth Objects since April 2014.

PS2 was built in an enclosure to the north of PS1, and connected to the same building as PS1. PS2 has a stiffer telescope that points better than PS1. PS2 has optics that should eventually deliver image quality slightly superior to PS1. PS2 began surveying the sky for NEOs in 2018. Since 2021, it has had similar performance to PS1.

The Pan-STARRS search for Near-Earth Objects is funded by the National Aeronautics and Space Administration (NASA) Near-Earth Observations Program, which is part of the NASA Planetary Defense Coordination Office (PDCO). The vast majority of NEO discoveries at the present time come from programs sponsored by the United States via NASA and the PDCO.

The main aim of the Pan-STARRS survey is to find larger asteroids or comets that may impact Earth in the future, so that steps can be taken to deflect the object, eliminating the impact. Smaller asteroids that may impact Earth with only a few days warning may also be discovered. For these case, if the object is large enough that damage or injury is possible, warnings can be issued.

2. The Pan-STARRS NEO survey

Pan-STARRS presently searches for NEOs by obtaining a sequence of four 45 s exposures spaced over a time period of approximately 1 hour. These observations are usually obtained using a w-band filter, which has a bandpass of 400–820 nm. This filter is designed to maximize sensitivity, with wavelengths beyond 820 nm blocked to eliminate the bright OH emission in the near-infrared. The w-band filter is used when the moon is down, or up and less than 60% illuminated. When a brighter moon is in the sky, an i-band filter is used, and for the three nights closest to full moon, a z-band filter is used. Most of the NEO discoveries are made using the w-band filter because the sensitivity is approximately 1 magnitude deeper than with the i-band filter and bright moonlight. Because of the wide-field nature of the Pan-STARRS telescopes, a moon avoidance angle of $40°$ is generally used, so observations near full moon are typically taken away from the ecliptic.

Observations are typically acquired in "chunks" which are groupings of approximately 20 adjacent pointings covering an area of approximately 120 deg^2. A chunk takes approximately 1 hour to complete. Chunks are 15 degrees wide (1 hour) in Right Ascension, with variable height in declination. As many as 10 chunks (or 1,200 deg^2) can be observed during a winter night with one telescope. The Galactic plane is generally avoided due to the high star density. Observations span the declination range $-50°$ to $+90°$. From the $+20°$ latitude of Maui, the north celestial pole is at the same elevation as $-50°$ declination. The Daniel K. Inouye Solar Telescope (DKIST) has been constructed to the southwest of Pan-STARRS, and its large enclosure obstructs parts of the southwest sky. This means that the southernmost declinations must be observed in the southeast as they rise or in the south. The southern obstruction of DKIST is less severe for PS2 since it is located north of PS1. In a typical lunation, most of the available sky can be surveyed at least once with each telescope, with some repeats possible in a lunation with good weather.

The images acquired by the cameras are transferred to computers at the University of Hawaii on the island of Oahu via a fast fiber optic-based network, where they are detrended, photometrically and astrometrically calibrated, and warped Magnier *et al.* (2020a), Magnier *et al.* (2020b). The primary technique presently being used to find moving objects uses pairwise image subtraction. Moving objects must be detected in a minimum of three of the four images. The moving object detection has a seeing-dependent lower limit of motion of approximately 0.05 deg/day, and an upper limit of 10 deg/day. Beyond 10 deg/day, there are too many false linkages producing for moving object candidates. Non-Gaussian noise from the CCDs produces many false detections. These false detections effectively limit the sensitivity of the survey; the number of false detections overwhelms the real detections if a faint detection limit is not imposed. As a result, some Near-Earth Objects go unreported by the moving object detection, but can later be manually measured and reported. Archival Pan-STARRS images are therefore rich in faint unreported asteroids, including unreported faint NEO detections.

A secondary NEO detection technique subtracts a deep stacked image from individual images, and searches for moving objects. A third technique is being developed which will use a static sky catalog and source detection in individual images. Each technique is hampered by noise and cosmetic defects in the detectors.

All moving objects have a "digest2" score calculated, which is a measure of how unusual the motion of the object is. Objects with digest2 scores greater than 50 are submitted to the Minor Planet Center (MPC), and rapidly processed. If they are not linked to a known object, or to an object already on the NEO Confirmation Page (NEOCP), they are posted to the NEOCP, and astronomers across the world attempt to obtain follow up observations, so that a preliminary orbit can be calculated.

NEOs usually, but not always, have unusual motion due to their proximity to Earth. NEOs can sometimes have motion that is typical of a main-belt asteroid, and those cases will have a low digest2 score and will not be posted to the NEOCP. NEOs that are headed directly towards Earth can also have very slow motion Wainscoat *et al.* (2022) that makes them difficult to recognize as moving objects. NEOs that are nearby may exhibit non-linear motion (curvature) across the sky due to the motion of the observer as the Earth rotates around its axis. The digest2 score does not consider curvature — instead it is based on the overall vector of motion of the object. At the present time, curvature must be recognized by an astronomer reviewing the moving objects, and the object forced onto the NEOCP.

All other moving objects are also submitted to the MPC, where they are either linked to known objects or stored in the Isolated Tracklet File. Moving objects are also screened for comets, which have a slightly extended appearance. Comet candidates are also reported to the MPC. Pan-STARRS presently discovers over 40% of all new comets.

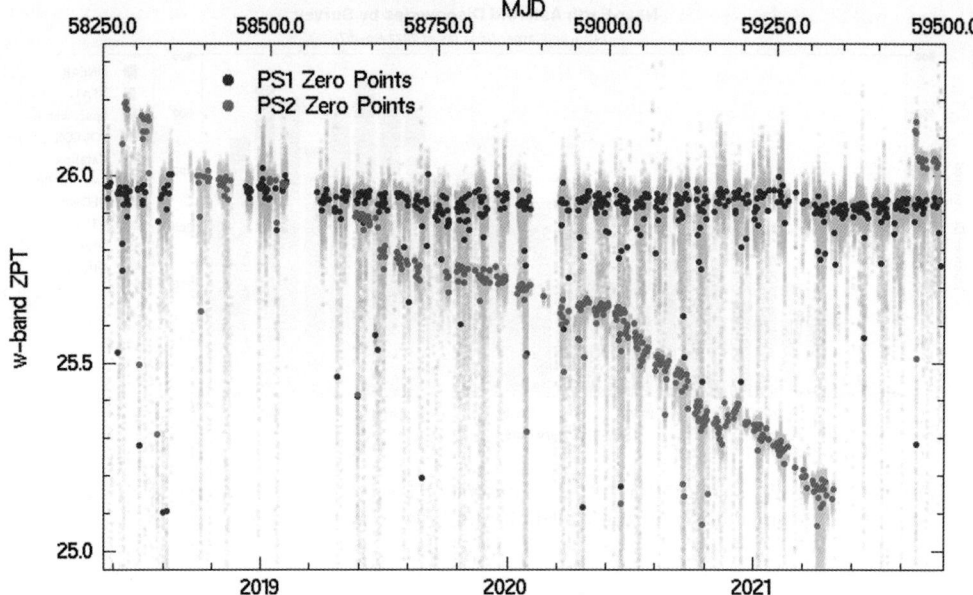

Figure 1. Photometric zero points for the Pan-STARRS1 and Pan-STARRS2 telescopes for 2018–2022.

3. Pan-STARRS2 secondary mirror problems

Despite optics that should have been better than PS1, PS2 lagged PS1 in NEO discoveries. This was initially attributed to poorer collimation and alignment, which was leading to poorer image quality. However, as the collimation and alignment was improved, the NEO discoveries by PS2 only increased modestly, and still lagged PS1.

Careful analysis of the zero points of the images showed that the throughput of PS2 was dropping quickly as shown in Figure 1. This was likely caused by the 2019 eruption of Kilauea on the island of Hawaii. The inversion layer kept the volcanic gases at lower altitude than Mauna Kea, but it is likely that some of these gases (notably sulfur dioxide) came to Haleakala where they produced serious degradation of the protected silver coating of the PS2 secondary mirror.

The PS2 mirror was removed for recoating and two excessively long screws were found to have been exerting pressure on the secondary mirror, producing astigmatism. This astigmatism was being partially removed by reshaping the surface of the primary mirror using actuators. Together, these non-ideal mirror shapes were contributing to poor image quality. Once the mirror was reinstalled, the astigmatism was gone from the secondary mirror and removed from the primary mirror, resulting in much improved image quality. The secondary mirror was coated with protected aluminum — the same coating used on the PS1 primary and secondary mirrors, and the PS2 primary mirror.

Now that the recoated secondary mirror has been reinstalled and collimated, discovery counts of NEOs from PS1 and PS2 are very similar, and the two telescopes have similar sensitivities.

4. NEO discovery statistics

The Pan-STARRS telescopes are the leading telescopes for discovery of larger (> 140 m diameter) NEOs, as shown in Figure 2. Pan-STARRS has a slightly larger aperture than the Catalina 1.5 meter telescope (observatory code G96) and better seeing conditions, making it able to see fainter NEO candidates.

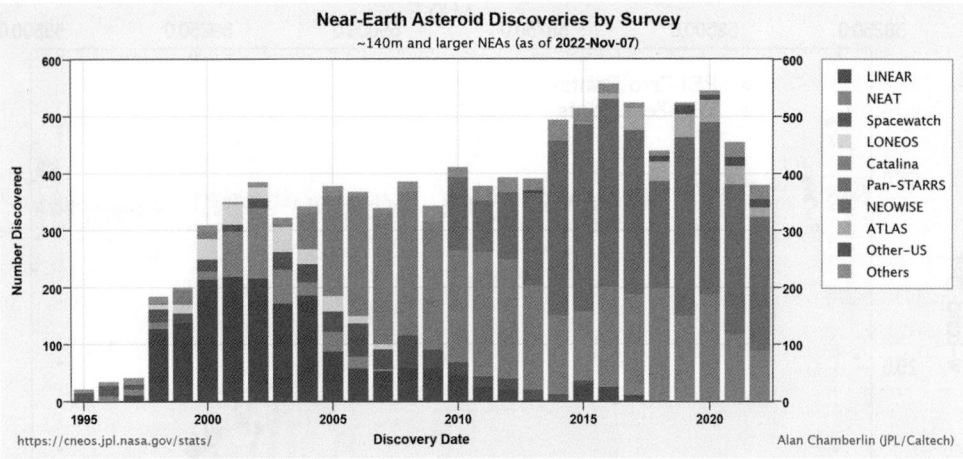

Figure 2. Discovery counts for 140-m diameter or larger NEOs. Pan-STARRS is colored magenta, and the Catalina Sky Survey is colored green.

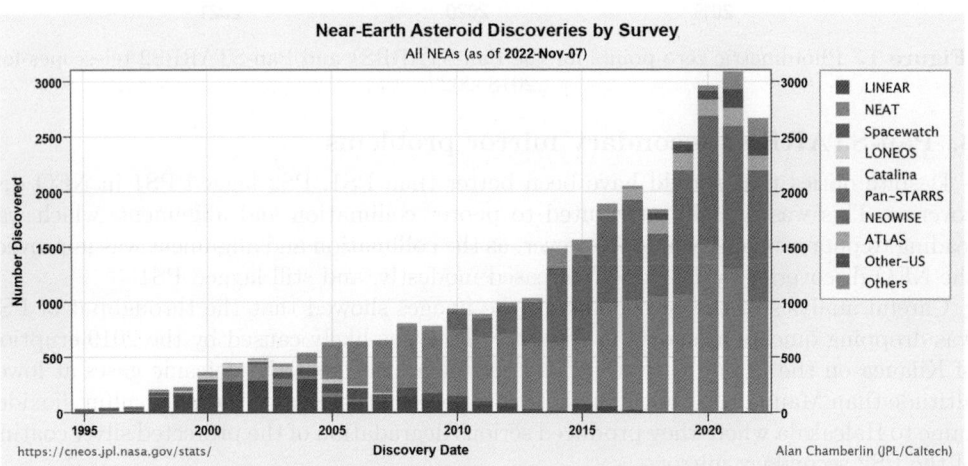

Figure 3. Discovery counts for all NEOs. Pan-STARRS is colored magenta, and the Catalina Sky Survey is colored green.

The Pan-STARRS telescopes are not as efficient at discovering smaller nearby NEOs and Atens because these faster moving objects often move into the cell gaps and their detections are corrupted. Figure 3 shows the discovery counts for all sizes of NEO, and shows that Pan-STARRS discovers a similar number of NEOs each year to the Catalina Sky Survey. Comparison of these two figures shows that the Catalina Sky Survey is more efficient at discovering smaller, faster moving NEOs.

4.1. *Reducing reporting latency and self-follow up*

For the period 2014–2020, most NEOs from Pan-STARRS were reported to the MPC on the morning following the observations. The basic rationale was that there were very few NEO follow up assets immediately to the west of Hawaii, so reporting NEOs in the morning HST was adequate for follow up from telescopes in Europe.

Uncertainties grow quickly for rapidly moving NEO candidates — particularly for NEO candidates moving to the east. The lack of rapid follow up contributed to the inefficiency of Pan-STARRS in discovery of smaller, faster-moving NEOs.

Beginning in 2021, Pan-STARRS modified its reporting so that most NEO candidates are reported within a few hours of observations. This has been achieved by having astronomers in the Hawaii time zone report NEO candidates until approximately midnight HST, and an astronomer located in a time zone located well to the east of Hawaii report NEO candidates in the morning in that time zone.

The rapid reporting has in turn allowed Pan-STARRS to perform self-follow up of its own NEO candidates. This self-follow up, usually in the form of a pair of observations, extend the arc from 1 hour to approximately 3–4 hours. This is powerful because Earth's rotation provides parallax, which in turn constrains the distance of the NEO candidate, and constrains the positional uncertainty, making future recovery easier.

Self-follow up is not possible for NEO candidates discovered late in the night because the morning twilight interferes. The rising waning gibbous moon can also interfere with follow up.

5. Future prospects

It is clear that NEO discoveries by Pan-STARRS could be improved by replacing the cameras. The OTA CCDs are not ideal for NEO discovery, due to their cell structure, and the OTA CCDs also have cosmetic problems. Large format commercially produced CCDs are available, but are costly. Pan-STARS would need 18 CCDs of the size that Catalina and ATLAS use in their telescopes. The cost of the CCDs alone would be ~\$5 million for each camera

New cameras could double the NEO discovery rate by Pan-STARRS. The increase in discoveries would come from increased quantum efficiency, improved cosmetics, reduced noise, larger area covered. together with elimination of the cell structure. Maintaining the present 10 micron pixels is advisable. 9 micron pixels (as used by ATLAS and Catalina) are too small; 12 micron pixels would be acceptable. Teledyne e2V makes a suitable CCD with 10 micron pixels. These CCDs are slightly rectangular in shape. The focal plane could be covered using 18 of these CCDs, with a top and bottom row of 4 CCDs, and two rows of 5 CCDs in the middle.

New CCDs would allow longer exposure times to be used. Exposure times at present in the w-band filter are limited to 45 seconds, because of the image persistence problem in PS1. New CCDs would not suffer from this image persistence, and so could be used for longer exposure times. Longer exposure times would help discover fainter slow-moving NEOs. Extending the NEO search to fainter magnitudes would allow for synergy with the NEO discovery that is expected from the Rubin Observatory in the south.

5.1. *Interstellar Objects*

In 2017, Pan-STARRS1 discovered the first interstellar object, 'Oumuamua. 'Oumuamua was discovered on a night with poor seeing conditions. Many fast-moving NEO candidates that Pan-STARRS reports are not recovered, and it seems likely that Pan-STARRS has seen other interstellar objects in the past that were not recognized as being interstellar.

Now that two Pan-STARRS telescopes are operating nightly (weather permitting), the likelihood of additional interstellar object discoveries increases. The rapid self-follow up that is being obtained for most NEO candidates will extend observational arcs to approximately 4 hours, and provide insight into whether a fast-moving object is nearby or more distant, with its fast apparent motion caused by high velocity rather than proximity to Earth.

5.2. *Impacting Objects*

Although the Catalina Sky Survey and ATLAS have discovered a few very small asteroids immediately before impact, Pan-STARRS has not discovered an impacting asteroid. Pan-STARRS may well have seen a few very small asteroids before impact, but they were not recognized as impacting objects, and follow up observations were not acquired. Now that Pan-STARRS is acquiring rapid same-night self-follow up for most objects, it is likely that imminent small impactors will be more easily recognized, and adequate follow up acquired before impact. And now that two telescopes are surveying the sky nightly (weather permitting), the likelihood of Pan-STARRS discovering an imminent impactor is doubled. Discovering small impacting asteroids is of great interest, since their astrometric measurements produce real-world insight that will be useful for larger, more dangerous impacts. Photometry and spectroscopy can yield the size of the object before impact, and if the impact occurs over land, meteorite samples can be compared to the spectral measurements acquired before impact.

References

Harris, A.W. and d'Abramo, G., 2015, Icarus 257, 302
Mainzer, A. et. al, 2011, ApJ, 743, 156
Magnier, E. et. al, 2020a, ApJS, 251, 3
Magnier, E. et. al, 2020b, ApJS, 251, 6
Wainscoat, R., et al., 2022 Icarus, 373, 114735

Astronomical Hazards for Life on Earth
Proceedings IAU Symposium No. 374, 2025
G. Tancredi, ed.
doi:10.1017/S1743921324000772

Optimization of Gauss Method to describe with most accuracy the orbits of Near Earth Asteroids - NEAs

Johana Murcia[1], Nicolás Molina[2], Miguel E. Gámez López[3], Néstor Méndez[1], Eduardo Mafla[4] and Camilo Delgado-Correal[5] ⓘ

[1]Department physics, National Pedagogical University

[2]Astronomical observatory, National University of Colombia

[3]Department physics, National University of Colombia

[4]Educational institution Compartir of Mosquera

[5]Faculty of Engineering, Distrital University FJC
email: mcdelgadoc@udistrital.edu.co

Abstract. In this work we reviewed the Gauss method to infer the orbits of minor bodies of the solar system, as the identification of optimization parameters to infer the orbital elements of two asteroids near to the Earth (NEAs): 5587 (1990 SB) and 4953 (1990 MU)), already cataloged and named by the Minor Planet Center - MPC. We used the database of JPL - Horizons and also included an analysis using data of Gaia Data Release 3 (DR3). We did an statistic analysis between distribution of points correspondent to orbital elements that we obtained and the inferred by the Jet Propulsion Laboratory-JPL. When data is crossed with the orbital parameters of Small-Body Database (JPL), we analyzed the deficiency grades of the method by statistic comparison with state vector method between orbital elements inferred with the implemented method and the accepted by the community, obtaining a relative error for the orbits calculated of 0.424149% and 0.416237% for the body 5587 (1990SB), and 0.257968% and 0.223521% for the body 4953 (1990MU). The first error value mentioned corresponds to an orbit calculated with the database JPL- Horizons, and the second value to an orbit calculated with the database Gaia (DR3), for each body in study. In the first phases of implementation of the code, it was found that the restrictions of the traditional method are overcome under the additional parameters that are proposed, resulting in orbits that are better approximated than those determined by NASA Team at the time of observation corresponding to the collection of data for the different bodies analyzed in the framework of this work.The best approximations are established with respect to the calculation of the orbital elements through the numerical solutions incorporated in the NASA SPICE kernel in python for a period of 100 years, with a step of 1 month. We found that our data are quite close to the curves that represent the variations of each of the 5 orbital elements involved in our analysis, namely: a, e, i, ω, Ω. Finally, it should be noted that our method could contribute to the estimation of orbits for minor bodies of the Solar System from observational data, which could easily be taken by using small telescopes. Thus, it would enrich processes that seek to expand the coverage of observatories focused on estimating the orbits of minor bodies in the solar system.

Keywords. Method Gauss, Asteroids, Optimization parameters

---

1. Introduction

A major problem in orbit determination is the calculation of the orbital elements of a celestial body using a given set of angular observations(α_i, δ_i)(Gronchi (2004)), for this, the Gaussian method is implemented (Gurfil & Seidelmann (2016)). The focus of the problem in this work is to find the distance from the Sun of two near-Earth asteroids (5587 (1990 SB) and 4953 (1990 MU)), knowing only their relative positions in the celestial vault to three observation times. For this reason, a possible optimization of the Gaussian method is proposed, taking into account **the interval of separation of observation dates**, because by setting this criterion well, a well-defined orbit can be obtained.

2. Gauss method for preliminary orbit inference

The Gauss's method for determining preliminary orbits is an iterative process and uses vector algebra for all his deductions.

We can see that between each set of vectors there is the following mathematical relationship: $\vec{r}_i = -\vec{R}_i + \vec{\rho}_i$, where $\vec{r}_i$ are the Asteroid position vectors with respect to the Sun, $\vec{R}_i$ represent the position of the Earth with respect to the sun and $\vec{\rho}_i$ is the position of the smaller body with respect to the Earth, with $i = 1, 2, 3$ corresponding to the indices of the radio vectors of each observation.

Notice that $\vec{\rho}_i$ is a geocentric vector, and since the observations are referred to the celestial sphere, this vector can be rewritten in terms of its components using spherical coordinates, thus:

$$\vec{\rho}_i = \rho_i(\cos\alpha_i \cos\delta_i \hat{i} + \sin\alpha_i \cos\delta_i \hat{j} + \sin\delta_i \hat{k}) = \rho_i \hat{u}_i \tag{2.1}$$

On the same hand, the following vector sum can be established between each of the vectors $\vec{r}_1$, $\vec{r}_2$ and $\vec{r}_3$ in the figure (1) given their coplanarity relationship.

$$\vec{r_2} = c_1 \vec{r_1} + c_3 \vec{r_3} \tag{2.2}$$

Substituting the equation (2.1) on the equation (2.2) and organizing the expression, we obtained:

$$\rho_2 \hat{u}_2 - c_1 \rho_1 \hat{u}_1 - c_3 \rho_3 \hat{u}_3 = \vec{R}_2 - c_1 \vec{R}_1 - c_3 \vec{R}_3 \tag{2.3}$$

Now the key point to Gauss's method is to find some value of ρ_i, in terms of what is already known, so we must eliminate ρ_i like ρ_1 and ρ_3 and find ρ_2. To do that, a vector must be found that is perpendicular to $\hat{u}_1$ and $\hat{u}_3$, this is the vector $(\hat{u}_1 \times \hat{u}_3)$, developing a vector elimination method and applying the corresponding algebra, the following general expression can be deduced:

$$\rho_j = \frac{-\vec{R}_j \cdot (\hat{u}_i \times \hat{u}_k) + c_i \vec{R}_i \cdot (\hat{u}_i \times \hat{u}_k) + c_k \vec{R}_k \cdot (\hat{u}_i \times \hat{u}_k)}{c_j \hat{u}_j \cdot (\hat{u}_i \times \hat{u}_k)} \tag{2.4}$$

With i, j, k different swapping between 1,2,3. The terms c_1 and c_3 that result from the equation (2.4), are obtained by means of the Lagrange coefficients which result from considering a vector function that can be expressed as a Taylor-type approximation to express a position vector which results from the analysis of two bodies between the Sun and the minor body in its second position.

$$c_1 = \frac{g_3}{f_1 g_3 - g_1 f_3} \qquad c_3 = \frac{-g_1}{f_1 g_3 - g_1 f_3}$$

Initially, the method assumes knowing the Lagrange coefficients f_i and g_i, that depend on $\vec{r}_2$ and its time derivative. Any value of $\vec{r}_2$ can be assumed, which will be progressively updated as the procedure iterates again, hoping to reach a convergence value of $\vec{r}_2$ given

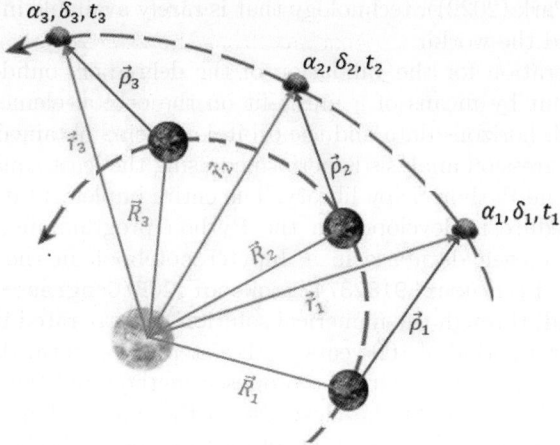

Figure 1. Graphical representation of the Sun-Earth-Asteroid position vectors. Image taken and modified from Zuluaga (2021)

a certain number of iterations, knowing rho_2 it is possible to find the real value of $\vec{r}_2$, applying the following equation:

$$\vec{r}_2 = \rho_2 \hat{u}_2 - \vec{R}_2 \tag{2.5}$$

Generally with $(i = 1, 2, 3)$. Once the position of the body has been determined with the equation (2.5), to obtain the orbital elements, it is necessary to know the speed of the smaller body, for this the following equation is used Curtis (2013):

$$\vec{\dot{r}}_2 = \frac{f_1 \vec{r}_3 - f_3 \vec{r}_1}{f_1 g_3 - f_3 g_1} \tag{2.6}$$

Finally, with the state vector that describes the movement of the asteroid, the orbital elements are calculated as have been done by Weber (s.f)

3. Gauss method optimization

From the previous section, it can be inferred that knowing the observation times with certainty is essential to successfully determine the orbit of an asteroid, that is the raison that a code is executed, for organized date ranges in the observations.

We work with two sets of observations. The first has samples of observations for each NEA's (5587 (1990 SB) and 4953 (1990 MU)), randomly distributed over time, provided by the GAIA probe; For each data sample, the values with the same observation dates are eliminated and permutation combinations are made between the resulting data in order to obtain the largest amount of observational data. The second set consists of two samples of observations ordered in time (each one corresponding to each body), separated by days, starting with intervals from 1 to 264. It is worth mentioning that the choice of the two bodies for which their orbit, is carried out by virtue of the quantity and separation between dates of observations made by the GAIA probe and its classification.

After inferring the possible orbits for each body and taking into account each database, the data obtained are compared with those calculated by NASA's JPL (Jet Propulsion Laboratory) and compiled in the MPC (Minor Planet Center), it is important to mention that JPL works with the construction of a state vector that is obtained by means of radar astrometry (Ryan Park (2022)), technology that is rarely available in many astronomical observatories around the world.

The contrast operation for the validation of the degree of confidence of the applied method is carried out by means of a linear fit on the orbital elements inferred for the GAIA data, the JPL horizons data and the orbital elements obtained for these bodies in the MPC. Linear regression analysis is performed using the least squares fitting method, with support from the Python Scipy library. The entire implementation, adaptation and data analysis procedure is developed in the Python programming language (†). The method is coded for each database in a Jupyter notebook in the github repository:: `https://gitfront.io/r/user-9182374/kmqWenqrt2CE/Congreso-IAU-2022/`.

On the other hand, through the numerical solutions incorporated in the NASA SPICE kernel in Python for a period of 100 years, with a step of 1 month, the degree of approximation of the elements inferred with the proposed method and the one used is verified. by JPL Horinzons, by means of the closeness to the curves that represent the variations of the orbital elements (a, e, i, ω, Ω) of each one of the bodies when subjected to perturbations, for this, the Sun-Earth-Moon-NEO system is modeled.

4. Results

The orbital elements are shown (delimited in a range of values according to the orbital characteristics of the NEA), the error calculated for the orbit obtained with each of the data sets and the one that meets the condition of being the most consistent with that determined by the JPL Horizons. For each of the bodies and with the results of each of the databases used, the first four sets of orbital element values are chosen. (a,e,i,ω,Ω) calculated, whose characteristic is approximately to the theoretical values determined by the MPC. Finally, the perturbed orbital elements are plotted, along with those inferred by each method. In total, 546 sets of observations (α,δ, t) are used for the first body, which is the love type, and 264 sets of observations (α,δ, t) for the second body, this being an apollo type under the category of potentially dangerous; the above data is for the GAIA probe.

At next, it is represented in histograms (figure 2 and figure 3), the distribution of points corresponding to the orbital elements calculated according to delta of randomly distributed observations for GAIA data and progressively systematized with deltas of increasing days with JPL Horizon. Note the number of calculated orbital element values that correspond and are close to those measured by the MPC for the epoch August 09, 2022 for both GAIA (left column) and JPL Horizon (right column) data.

From these represented data, for the orbits of the two minor bodies in question, the list of observations whose total orbital elements are closer to those measured by the MPC are filtered. In total, the best four orbits with the lowest standard deviations with respect to the values reported by the MPC are delimited. There are four orbits for each data set (GAIA, JPL) of each body. These new orbital element datasets are plotted as a linear regression, in order to estimate a global correlation coefficient between data measured by the MPC and those calculated for each data source (GAIA and JPL) for each body.

The expected value of said coefficient is 1, and corresponds to the correlation 1.1 for each of the MPC orbital elements and calculated according to each source. The ordering of the values is given automatically, and they are represented on a logarithmic scale. The correlation coefficients for each data set are arranged in the table 1.

Finally, from the four orbits selected for each body with each database, the one that is closest to the one calculated by the MPC is chosen. The results of the orbital elements that define the best orbit calculated with each database, are shown in the table 2.

† PythonSoftwareFoundation (2022). `https://www.python.org`

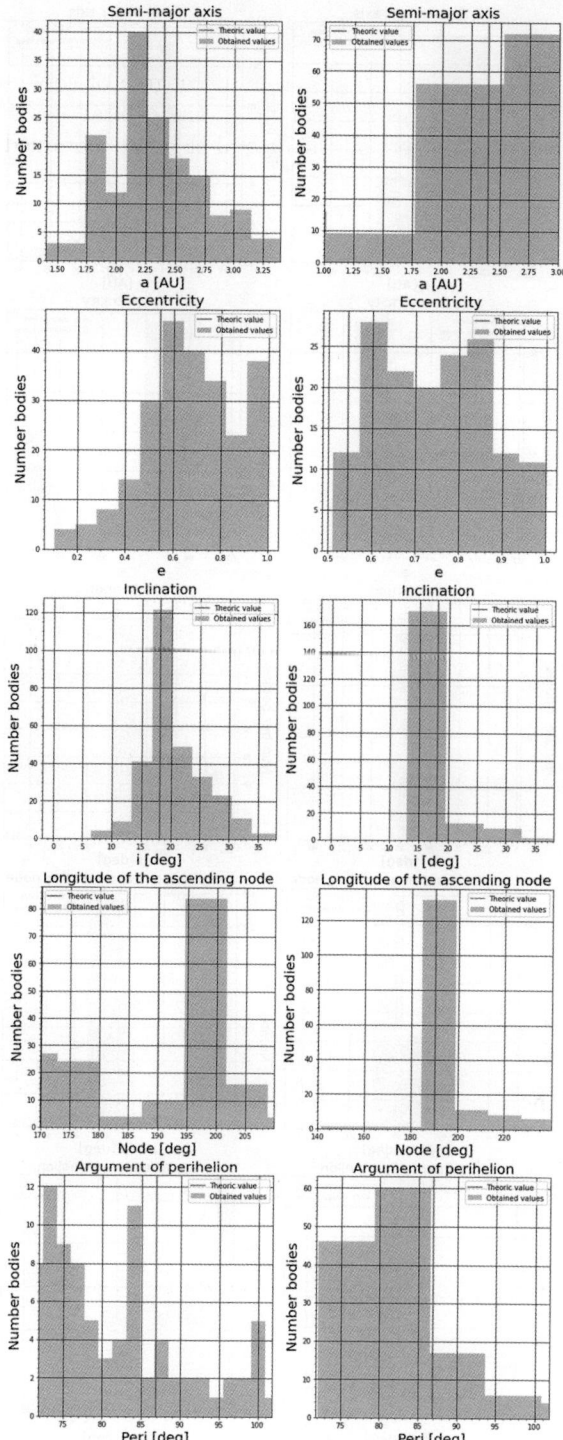

Figure 2. Histograms of orbital data for body 5587 (1990 SB), obtained with data from the GAIA probe and JPL Horinzons.

J. Murcia *et al.*

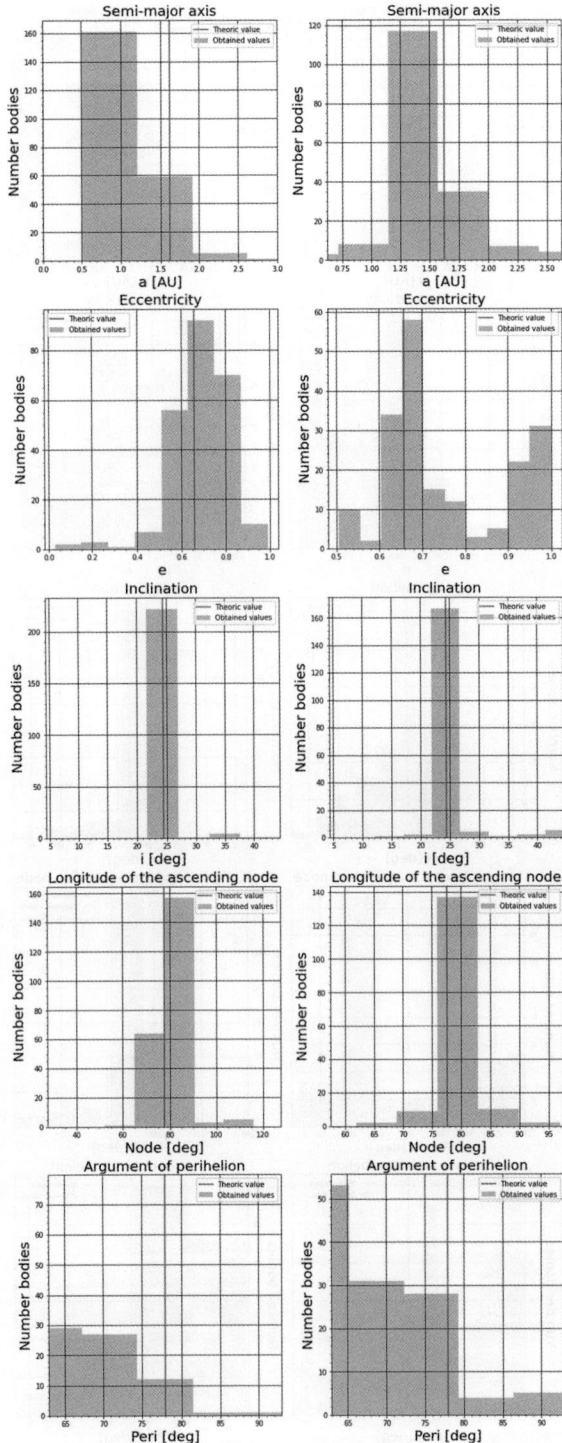

Figure 3. Histograms of orbital data for body 4953 (1990 MU), obtained with data from the GAIA probe and JPL Horinzons.

Table 1. Correlation coefficients for bodies 5587 (1990 SB) and 4953 (1990 MU)

Name of NEA	Database	Slope	R^2
5587 (1990SB)	GAIA	0.852	0.906
5587 (1990SB)	JPL Horizons	0.993	0.997
4953 (1990 MU)	GAIA	0.997	0.999
4953 (1990 MU)	JPL Horizons	1.013	0.994

Table 2. Orbital elements obtained for the best orbit of the bodies 5587 (1990 SB) and 4953 (1990 MU) with each of the databases (JPL-Horizons and Gaia), compared to those determined by the MPC

Name of NEA	Database	a (AU)	e	i (deg)	Node (deg)	Peri (deg)
5587(1990 SB)	MPC	2.3976	0.5441	18.0946	189.9204	86.8275
5587(1990 SB)	JPL- Horizons	2.1953	0.5528	18.1252	196.1242	86.1973
5587(1990 SB)	Gaia	2.2858	0.5935	18.0893	196.1899	75.0662
4953 (1990 MU)	MPC	1.6216	0.6574	24.3739	77.5534	77.9210
4953 (1990 MU)	JPL- Horizons	1.3721	0.6512	24.3398	77.7148	77.9633
4953 (1990 MU)	Gaia	1.3922	0.6333	23.9818	73.4068	76.9167

Table 3. Error calculation of the computed orbits for bodies 5587 (1990 SB) and 4953 (1990 MU)

Name of NEA	Database	Error (%)
5587 (1990SB)	JPL Horizons	0.424149
5587 (1990SB)	GAIA	0.416237
4953 (1990 MU)	JPL Horizons	0.257968
4953 (1990 MU)	GAIA	0.223521

On the table 3, The error for each calculated orbit is presented, with respect to the measurements made by MPC for the year 2022.

In the context of the results previously obtained, a comparison was established between the orbital elements derived from the two NEOs studied, from the GAIA and JPL Horizons ephemeris databases, and the numerical solutions reported by the temporal evolution of the vector of state obtained using the NASA SPICE library, which incorporates second-order perturbations for the orbital elements of the bodies studied in an N-body astrodynamic problem. The program was executed for a time interval of 100 years with deltas corresponding to 1 month, and later the temporal evolution of the state vector of each body was transformed into functions that showed variability in their orbital elements. Contrasting the temporal evolution of each one of the orbital elements of the two studied NEOs, with the orbital elements inferred according to the two different data sources (GAIA and JPL Horizons), and the elements orbitals calculated for the corresponding epoch reported by the MPC through JPL, we obtained the figure 4.

It can be notice that, in general, the proposed optimization method produces orbital elements that are better approximated than those reported by MPC, while working with ephemeris data from JPL Horizons. "Best Approximation" refers to the optimization method takes values for each orbital element, which intersect with the values of the corresponding orbital elements, in the curves produced by numerical computation of state vectors in an N-body problem.

5. Conclusions

The Gauss method proposes that data collection should be equally spaced between observations, however, it is evident that although the GAIA data do not get this condition and are difficult to link to the method. It is possible to obtain approximate orbits to the

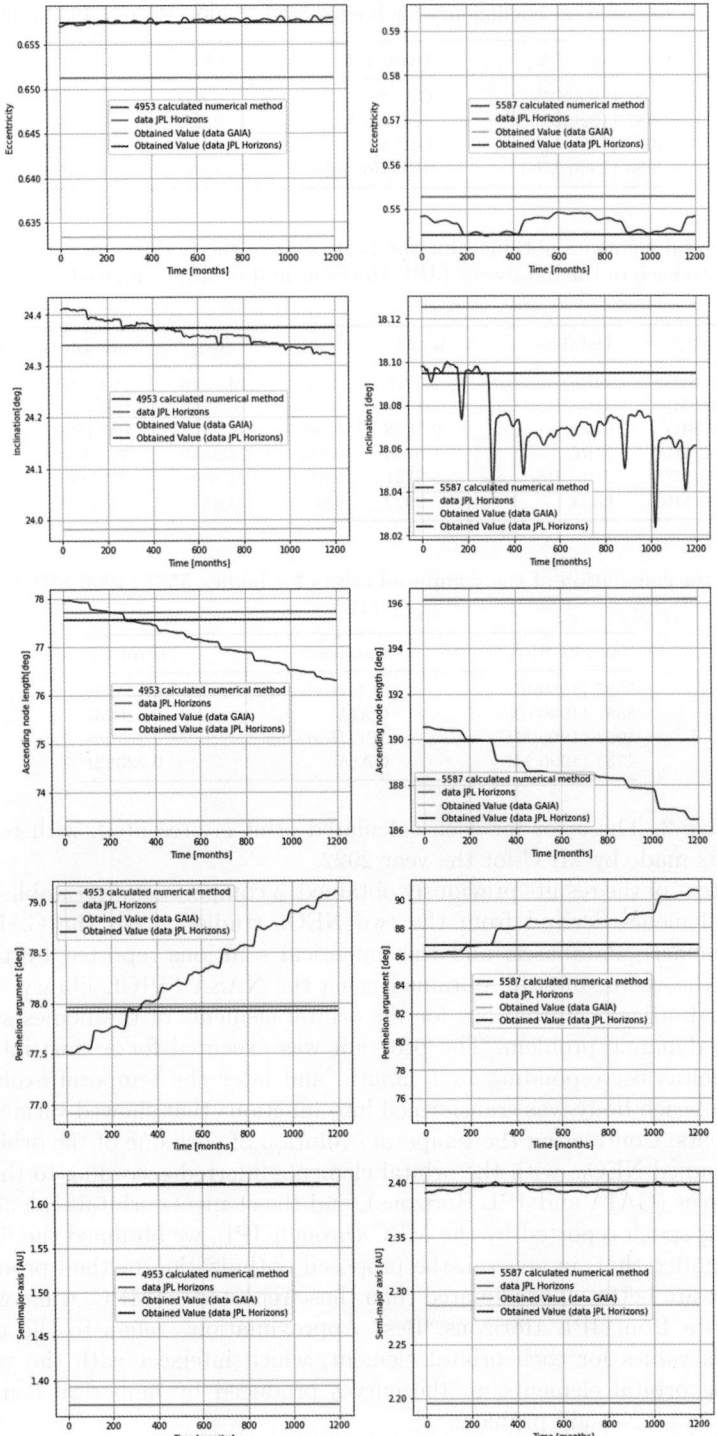

Figure 4. Temporal variability of orbital elements of the two analyzed NEOs. The data in the left column correspond to 4953 (1990 MU), and those in the right column to 5587 (1990 SB).

resulting ones with the JPL Horizons data and those calculated by the MPC, under the implementation of the corrections proposed in this study.

The proposed method in this work can be incorporated to optimize the observation in small observatories, since only 3 observations are needed without the necessity of radar technology or very high resolution telescopes, to calculate preliminary orbits of small bodies whose approximation is consistent with the data calculated by the MPC.

The obtained results for the calculation of the orbital elements through the optimization method incorporated here, in contrast to the results reported by the MPC for the last corresponding period of observation (August 2022), bring up the problem of establishing with respect to what sets of orbital elements we are evaluating the error in the inference of the orbits. The error results in the comparison of the different derivations, with what is established by the theoretical models that predict by means of numerical methods the way in which the orbital elements of some small body in the solar system varied in an N-body problem. If this is so, the question remains open about the criteria that prioritize theoretical derivations over empirical estimates made with observations both by MPC, as well as by our method, or any other that is incorporated in this direction.

Taking as reference the curves that produce the solution of the N-body orbital problem translated into the temporal evolution of each one of the orbital elements, our optimization method for JPL data suggests better approximate results than those reported by the MPC, and those calculated in our method with GAIA data.

References

Curtis, H. D. (2013). *Orbital mechanics for engineering students*. Butterworth-Heinemann.

Gronchi, G. F. (2004). Classical and modern orbit determination for asteroids. *Proceedings of the International Astronomical Union, 2004* (IAUC196), 293–303.

Gurfil, P., & Seidelmann, P. K. (2016). *Celestial mechanics and astrodynamics: theory and practice*, vol. 436. Springer.

Ryan Park, A. B. C. (2022). Small-body radar astrometry.
URL https://ssd.jpl.nasa.gov/sb/radar.html

Weber, B. (s.f). Orbital mechanics & astrodynamics.
URL https://orbital-mechanics.space/classical-orbital-elements/orbital-elements-and-the-state-vector.html

Zuluaga, J. (2021). Sb - especial - método de gauss de determinación de órbitas.
URL https://www.youtube.com/watch?v=ri8Zf_LvOsI

Astronomical Hazards for Life on Earth
Proceedings IAU Symposium No. 374, 2025
G. Tancredi, ed.
doi:10.1017/S174392132400067X

Piggybacking astronomical hazard investigations on scientific Big Data missions

Gijs A. Verdoes Kleijn[1,2] ⓘ, Teymoor Saifollahi[1], Rees Williams[3], Oscar Stolk[1] and Georg Feulner[4]

[1]Kapteyn Astronomical Institute, University of Groningen, The Netherlands
email: `g.a.verdoes.kleijn@rug.nl`

[2]Netherlands Research School for Astronomy, The Netherlands

[3]Donald Smits Centre for Information Technology, University of Groningen, The Netherlands

[4]Potsdam Institute for Climate Impact Research, Germany

Abstract. Current and upcoming large optical and near-infrared astronomical surveys have fundamental science as their primary drivers. To cater to those, these missions scan large fractions of the entire sky at multiple wavelengths and epochs. These aspects make these data sets also valuable for investigations into astronomical hazards for life on Earth. The Netherlands Research School for Astronomy (NOVA) is a partner in several optical / near-infrared surveys. In this paper we focus on the astronomical hazard value for two sets of those: the surveys with the OmegaCAM wide-field imager at the VST and with the Euclid Mission. For each of them we provide a brief overview of the astronomical survey hardware, the data and the information systems. We present first results related to the astronomical hazard investigations. We evaluate to what extent the existing functionality of the information systems covers the needs for the astronomical hazard investigations.

Keywords. minor planets, asteroids, comets, stellar proper motions, climate change, astrometry, surveys, Big Data, Data Science, information systems

1. Introduction

In the last four decades there has been an exponential growth in the observational data gathered by optical and near-infrared astronomical imaging surveys (see e.g., Fig. 1, Tyson 2019, Verdoes Kleijn 2023). This growth is continueing unabated. The Netherlands Research School for Astronomy (NOVA†) is partner in several optical / near-infrared surveys. In this paper we focus on two sets of those: the surveys with the OmegaCAM wide-field imager at the VST, in particular the Kilo-Degree Survey and the ground-based and space-based surveys part of the Euclid Mission. These missions have fundamental astronomical science as their primary driver. For this they perform observations of large fractions of the entire sky at multiple wavelengths and at multiple epochs. These aspects make these data sets also valuable for investigations into astronomical hazards for life on Earth. These survey missions will reach the tens of Terabytes regime in terms of catalogs and metadata databases and up to tens of Petabyte regime in terms of bulk data volume. This volume is spread over up to hundreds of thousands of exposures with each of them producing rich sets of metadata, such as catalogs. These in turn lead to millions of bulk data files. Therefore, the scientific exploitation and mining of these "Big Data" sets requires information systems which have an advanced databasing system at their

† `https://nova-astronomy.nl/`

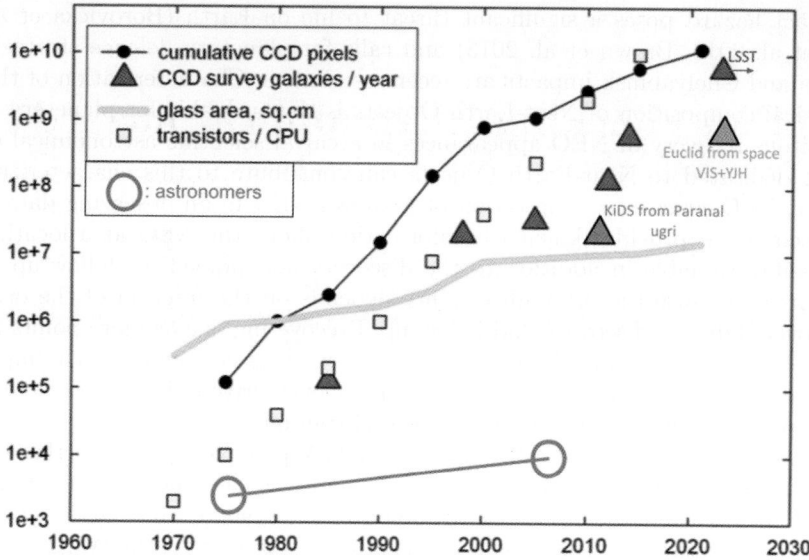

Figure 1. Growth of information technology, galaxy surveys and astronomers as a function of time. The typical number of transistors per CPU, the aperture area of telescopes ("glass"), the cumulative number of pixels in telescope cameras and number of astronomers are shown. The plot is taken from Verdoes Kleijn (2023) which adapted it from Tyson (2019). The triangles show for surveys the total number of galaxies observed with S/N > 5 for flux inside a 2 arcsec wide aperture per year in at least one survey filter. The arrow for the LSST triangle indicates it is currently planned to start late 2024.

core. Furthermore, to support the calibration of the raw data and its subsequent scientific analysis these systems also need to interface to high performance compute clusters and massive storage systems. For both survey missions NOVA has a leading role in the development and operation of the associated information systems. This offers NOVA a good opportunity to (re-)use both the observational data and the associated information systems for investigations into astronomical hazards, such as posed by asteroids, comets and close stellar encounters. This way NOVA can make a contribution to protecting society against astronomical hazards, on short and long terms. In this paper we describe two recently initiated pilot projects within NOVA investigating astronomical hazards that piggyback on the survey data and associated information systems developed for fundamental science. One project pertains to Near-Earth Objects using OmegaCAM data with the ASTROWISE information system. The other project pertains to climate change due to comet impacts due to close stellar encounters. This will use the surveys of the Euclid Mission and its Euclid Data Processing System. For both each we present first results related to the astronomical hazards. We also provide an overview of the astronomical survey hardware, the data and the information systems. In this we highlight to what extent the existing functionality of the information systems covers the needs for the astronomical hazard investigations.

2. Near-Earth Object precovery and discovery with OmegaCAM and AstroWISE

Near-Earth Objects (NEOs) are asteroids or comets whose perihelion occurs at less than 1.3 astronomical units (au), meaning that close approaches with the Earth might occur at some point. The size of these objects ranges from meters to tens of kilometers.

The impact hazard poses a significant threat to life on Earth (Borovička et al. 2013, Popova et al. 2013, Brown et al. 2013) and calls for planetary defence strategies. The Tunguska and Chelyabinsk impacts are recent reminders. Characterization of the orbits and physical composition of Near-Earth Objects is thus valuable for planetary defence. Serendipitous recovery of NEO appearances in archival scientific astronomical observations not dedicated to Near-Earth Objects can contribute to this characterization. In particular, NEO *precovery* − detection of a known NEO in an observing dataset prior to its discovery − provides kinematic information about the NEO at a location in its orbit possibly valuable in addition to the discovery and immediate follow-up observations. This is because the orbit uncertainty depends on the fraction of the orbital arc that is covered during discovery and follow-up. Precovering one or more points far away from the discovery and immediate follow-up observations can significantly improve the accuracy of the orbital parameters. In this way, science-driven missions not only have a scientific purpose but can also provide a societal spin-off.

Therefore we initiated with ESA and within NOVA an exploratory pilot that evaluates both the re-use of astronomical imaging surveys, in this case those of the OmegaCAM wide-field imager, and the re-use of the information system which was developed to handle the production and scientific analysis of these surveys: AstroWISE.

OmegaCAM is a wide-field camera on the VLT Survey Telescope at ESO's Cerro Paranal Observatory. OmegaCAM has 32 science CCDs with a field of view of approximately 1 square degree. Over the first decade of its operation, OmegaCAM has covered a significant portion of the southern hemisphere (Fig 2) in over 400 000 exposures.

AstroWISE stands for Astronomical Wide-field Imaging System for Europe which is an information system for data management, image processing, and calibration for a range of astronomical telescopes and instruments in a single data flow environment (Begeman et al. 2013, McFarland et al. 2013a). It has been used to do survey production for example for OmegaCAM's Kilo-Degree Survey (Kuijken et al. 2019) and the Fornax Deep Survey (Peletier et al. 2020).

For the pilot we developed an AstroWISE Precovery Pipeline which re-used the AstroWISE functionality for data calibration, processing and analysis and added automated interfaces to webservices for ephemerides prediction (SSOIS, Gwyn et al. 2012, JPL Horizons†) and includes the deployment of dedicated software for the detection of streaks (StreakDet, Virtanen et al. 2016, Pöntinen et al. 2020).

The pilot with the AstroWISE Precovery Pipeline resulted in the recovery of 196 appearances of NEOs from a set of 968 appearances predicted to be recoverable. The achieved astrometric and photometric accuracy is on average 0.12 arcsec and 0.1 mag. It includes 49 appearances from a set of 68 NEOs predicted to be recoverable and which were on ESA's and NASA's risk list at that point. ESA's risklist‡ is provided by the ESA near-Earth Objects Coordination Centre (ESA-NEOCC) and consists of known NEOs with a non-negligible chance of impact in the next hundred years. The appearances of three NEOs constituted precoveries, i.e., appearances well before their discovery. The subsequent risk assessment using the extracted astrometry removed these NEOs from the ESA and NASA risk list. For an in-depth discussion of the methods and results we refer to Saifollahi (2023).

Using the experience of the pilot we attempt here to answer questions on the value and challenges of re-using the astronomical data and associated information systems for planetary defence against NEOs.

---

† https://ssd.jpl.nasa.gov/horizons/app.html
‡ https://neo.ssa.esa.int/risk-list

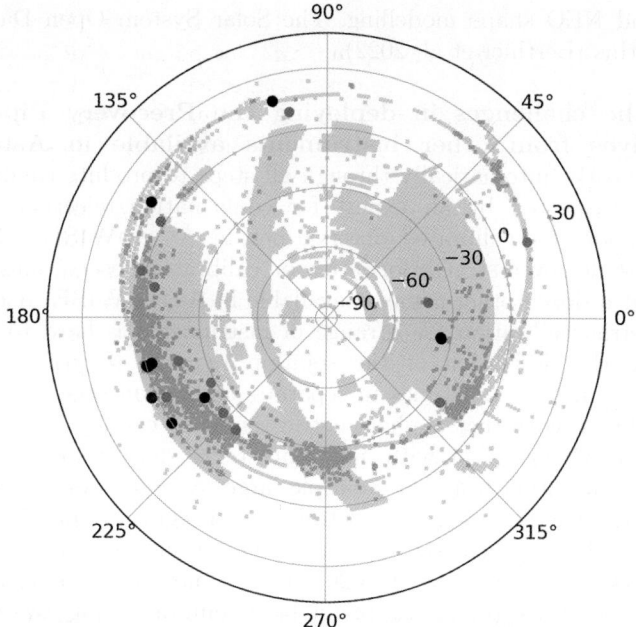

Figure 2. The sky coverage of the over 400 000 OmegaCAM/VST observations (green area). Of those about 10 000 frames are predicted to overlap with a known NEO (light blue dots). For the small subset predicted to have sufficient signal-to-noise, those with a successful and failed detection are shown as black and red points, respectively.

What is the detectability of NEO appearances in astronomical archives such as the OmegaCAM archive? We define NEO detectability as the fraction of detectable appearances among the total of occurrences that a NEO is predicted to be located within the FoV of the images. For the OmegaCAM pilot, the detectability varies as a function of the chosen threshold of signal-to-noise (SNR). The detectability rate is estimated to be ~0.005 at an SNR> 3 for NEOs on the risk-list and for the full list of NEOs. We expect no significant improvement can be made in detectability given the low 3σ threshold in predicted SNR and the fact that detected NEOs tend to be often a few tenths of magnitude fainter than predicted for this pilot.

What is the precovery rate for NEOs predicted to be detectable? The precovery rate for SNR>3 is 40% for NEOs on the risk-list and 20% for the full list of NEOs. The precovery rate increases to about 50% for SNR>10. So a factor of up to 5 more NEOs can be precovered from the OmegaCAM archive through improved detection techniques (see below for discussion on new techniques). It will be a significant result as well if, after improving the recovery processes, the failed recoveries turn out to be in fact non-appearances. It would suggest that the actual orbital accuracy for those objects (including those on the risk list) is significantly worse than predicted.

What astrometric and photometric accuracy can be achieved? The astrometric and photometric accuracies are 0.12arcsec (15% of the average FWHM of OmegaCAM/VST images of about 0.8arcsec) and 0.1 mag. Improvements in astrometric accuracy are expected from propagating the proper motions in the Gaia astrometric reference catalog to the observation date of the science image. Improvements in photometric accuracy can come from a more sophisticated modelling of SED, observational configuration and NEO shape modelling. The Solar System Open Database Network might facilitate this (Berthier et al. 2022).

What are the challenges in deploying the Precovery Pipeline, also on imaging archives from other instruments available in AstroWISE? The Pipeline works mostly automatically through all steps to produce candidate recoveries. These are then inspected by an expert for confirmation/rejection. Thanks to the common data model for calibrated observations in AstroWISE (McFarland et al. 2013b) it could be deployed straightforwardly on calibrated observations for the imaging archives of about a dozen other cameras available in AstroWISE. A challenge is that precise photometric calibration for a range of instruments is hard to fully automate. This is because the derivation of the solution requires reference stars sometimes inside the science images, sometimes in separate calibration observations. A potential solution would be to construct a photometric reference catalog that spans the entire sky observable by OmegaCAM with sufficient stellar density. This appears possible by aggregating information from the multiple large-scale surveys of the Southern Sky. Another main challenge is robust NEO detection and segmentation. This is also a main reason behind the the obtained recovery rates. StreakDet is a great tool for detecting high SNR streaks with sizes between 5-20 arcsec. However, its performance drops for faint and long streaks. Deep learning might be a solution to improve streak detection and ultimately NEO precovery (e.g., Pöntinen et al. 2020).

For an in-depth discussion about the recovery results for OmegaCAM and the feasibility of deploying it to other instruments, we refer the reader to Saifollahi (2023).

3. Close stellar encounters with Euclid surveys and systems

Close encounters of stars to the Sun can affect climate and life on Earth. The ionizing radiation and cosmic rays from supernovae could have a significant impact on both for encounters within 10 pc (Thomas, these proceedings). Stellar encounters within 1 pc can cause significant gravitational perturbations in our Solar System's Oort Cloud. These can lead to increased influx of comets and hence planetary impacts in the inner Solar System (Bailer-Jones, these proceedings). Close stellar encounters can also bring an increase in the influx of exocomets. They can originate in either the Oort cloud of the passing star or in the cloud's tidal streams (Portegies Zwart 2021). Impacts by comets and asteroids may have caused climate changes in the past (Brugger et al. 2017) and might do so again in the future.

Identifying close stellar encounters requires six dimensional phase space coordinates (three positions, three velocities) for stars in the Milky Way. The Gaia Mission has brought an enormous information leap in stellar phase space measurements. Its third data release provides an all sky astrometric reference frame sampled with almost 1.5 billion point sources down to 22nd magnitude (Gaia Collaboration et al. 2022). It also provides six-dimensional phase space coordinates for over 33 million stars down to G=14 with positions accurate at the milli-arcsecond (mas) level, proper motions at the mas / year level and radial velocities at the km/s level. This stellar sample allowed identification of 42 stars with encounters within 1 pc with a perihelion time up to 6 Myr in the past and future (Bailer-Jones 2022). From a similar analysis on Gaia's second data release it was estimated that about 15% of all close stellar encounters within 5 pc and within 6 Myr were detected at that point (Bailer-Jones et al. 2018). The associated inferred rate of encounters within 1 pc is about 20 per million year. The final Gaia release might roughly double the completeness and be able to detect encounters with perihelion times of order 10 Myr in past and future.

To increase this completeness level and perihelion time span one has to identify close encounters from stars fainter than observable by Gaia using deeper surveys. Five dimensions of the six-dimensional phase space can be obtained by combining imaging surveys observing the same sky area at multiple wavelengths and at multiple epochs. The multiple epochs allow to derive proper motions (in addition to the positions). The multiple wavelengths allow to derive photometric distances (see e.g., Chapter 3 in Speagle (2020)). ESA's Euclid Mission brings together space-based and ground-based surveys at multiple epochs and multiple wavelengths over almost 15 000 square degrees of sky. Observations at 9 wavelengths are gathered via 8 instruments, located at 7 telescopes in space and on the ground. The first observations which are now being re-used as part of the Euclid Mission occurred from the ground in August 2013. The space-based observations for the Euclid Mission will be obtained with ESA's Euclid satellite. It will be launched in July 2023 and observe the almost 15 000 square degrees of extragalactic sky in about 6 years (Laureijs et al. 2011, Euclid Collaboration et al. 2022). The ground-based observations are planned to be completed well before July 2029. The Euclid satellite will survey the extragalactic sky using a 1.2m telescope with two imagers. The visible imager (VIS) and Near Infrared Spectrometer and Photometer (NISP), sharing a 0.53 square degree Field of View. VIS will detect point sources down to a limiting magnitude of 25 (AB, 10σ for a point source measured using a 2 arcsec diameter aperture) using a very broad filter (550–900 nm). The Near-infrared Spectrograph and Photometer (NISP) will measure their photometry through Y, J, and H filters down to a magnitude limit of 23.5 (using same definition as VIS). All space observations of a sky area will be done at a single epoch. This data will be combined with data from ground-based telescopes in the optical filter u, g, r, i, z to matching depth. In the Northern hemisphere this will be with four surveys. The Canada-France Imaging Survey (CFIS, Ibata et al. 2017) observes in u and r. CFIS observations started in the first semester of 2015 and are done to full depth in a single epoch. The Waterloo Hawaii IfA G-band Survey (WHIGS†) observes in g. WHIGS started approximately 2022 and observes to full depth in a single epoch. The Panoramic Survey Telescope And Rapid Response Systems 1 and 2 (Pan-STARRS 1 & 2, Kaiser et al. 2010) observes in i band. Pan-STARRS observations cover the Euclid survey area since 2010, building up depth by many revisits over years until a few years after 2023. The Wide Imaging with Subaru HSC of the Euclid Sky (WHISHES‡) survey observes in z. It observes since the second semester of 2020 to full depth in single epochs. In the Southern hemisphere the Euclid survey area is covered by the Dark Energy survey in g, r, i and z (DES, Abbott et al. 2021) and the Large Survey of Space and Time (LSST, Ivezić et al. 2019) in u, g, r, i and z. DES observed from August 2013 until January 2019, building up depth in yearly revisits. The Vera Rubin Observatory plans to start the LSST survey late 2024 and has 10 years of planned operations. It will build up depth through many visits over many years.

Combining such a heterogeneous set of epochs and filters into a homogeneously set of order billion stellar positions, proper motions and distances requires a careful calibration approach using a information system that also allows ample quality control. The astrometric calibration might best be done via calibration against with zero proper motion and well-defined and consistent centroids across the optical and near-IR. For this reason Tian et al. (2017) used compact galaxies as calibrators. They combined Gaia (Data Release 1) with data from the SDSS, 2MASS and PanSTARRS surveys to obtain proper motions for 350 million sources with a characteristic systematic error of less than 0.3 mas/year and a typical precision of 1.5–2.0 mas/year. The Euclid survey area will contain of order a billion stars. For Euclid such an astrometric calibration effort can be

---

† https://www.skysurvey.cc/aboutus/
‡ https://www.skysurvey.cc/aboutus/

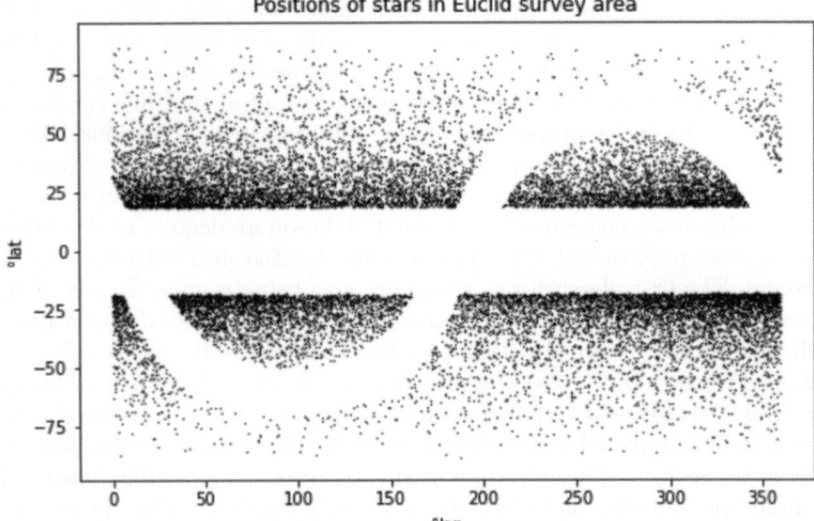

Figure 3. The position of stars in a simulated Milky Way distribution in Galactic coordinates. An Euclid's primary science driver is galaxy weak lensing tomography. Therefore its survey area is at galactic latitudes larger than —23 deg— to avoid the Milky Way disk and at ecliptic latitudes larger than —10 deg— to avoid near-infrared background contamination by zodiacal light. The simulation is made using the Galaxia modeling code (Sharma et al. 2011) in its wrapper code (Rybizki et al. 2018)

performed and released multiple times as observations on space and ground progress. To do such a massive operation repeatedly that accurately for so many objects is facilitated by an advanced databasing system providing a rich and detailed description of all data items (Mulder et al. 2020). The Euclid information system (called the Euclid Archive System) is such a "data-centric" system. It consists of two main components: the Euclid Science Archive System and the Euclid Data Processing System (Nieto et al. 2019). All bulk and metadata of calibration and science observations required for the re-use to determine close stellar encounters reside in the Data Processing System. Only a subset resides in the Science Archive System. For example, all bulk data, metadata and data quality reports related to individual ground-based exposures resides only in the Data Processing System. Combining Euclid's five-dimensional phase space (positions, distances and proper motions) with stellar radial velocities, from e.g., a spectral survey, establishes then finally the six-dimensional data set from which one can infer which of these billion stars (mostly fainter than observable by Gaia) lead/led to close stellar encounters.

Simulations are on going to determine what completeness and accuracy to expect in terms of nearby stellar encounters using Euclid Mission's observations gathered over almost two decades at 9 wavelengths when combined with such a radial velocity survey. Fig 3 shows a simulated Milky Way stellar distribution as observed by Euclid. The simulation is made using the Galaxia modeling code (Sharma et al. 2011) in its wrapper code (Rybizki et al. 2018). The code simulates magnitudes in all 9 Euclid filters and 6D phase coordinates. Photometric distance estimates can be derived using the magnitudes as input to a Bayesian statistical framework and modeling (e.g., Speagle 2020). Proper motions could be obtained from the positional catalogs of all observations following the approach of Tian et al. (2017). Finally stellar radial velocities have to be supplied by another mission than Euclid. The perihelion distance and time of close stellar encounters

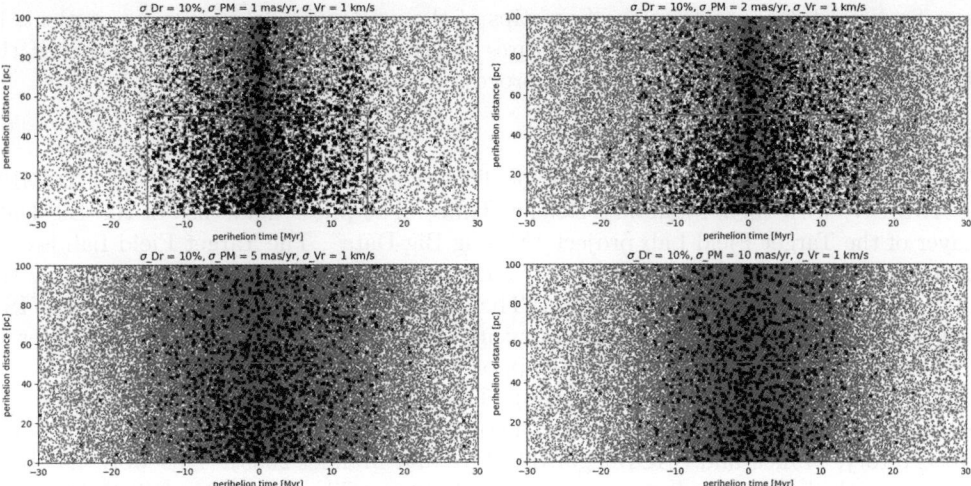

Figure 4. Estimated perihelion distances versus perihelion time for the stars in a simulated Milky Way distribution that can be detected by Euclid. This preliminary simulation selected randomly just 1% of the simulated Milky Way stars before performing the linear motion approximation to estimate perihelion time and distance. The small blue dots show ten perihelion estimates for each simulated star by sampling ten times independent Gaussian random errors in distance (σ_Dr), 2D proper motion (σ_PM) and radial velocity (σ_Vr). The large black dots show the true perihelion distance and time for stars that have an estimated perihelion to lie inside the red box of encounters within 50 pc within 15 million years. Completeness and false positive rate in a sample of candidate close encounters clearly depend critically on the measurement errors.

can then be estimated using the Linear Motion Approximation (Bailer-Jones et al. 2018). Very preliminary estimates are shown in Fig 4 for four different assumptions on accuracy of estimated distances, proper motions and radial velocities. Completeness and false positive rate in a sample of candidate close encounters clearly depend critically on the measurement errors.

4. Lessons learned

Above's two pilots have shown to us that it is worthwhile to explore further the piggybacking of astronomical hazard investigations onto the data and information systems developed for astronomical scientific Big Data missions. During the execution of above's two pilots we noticed two general characteristics of the astronomical scientific information systems that are key in making them more amenable for re-use in astronomical hazard investigations:

• Standardization is important. Using common interfaces, data models and data structures inside a single information system / environment for the datasets from different instruments makes it significantly less effort to develop and implement pipelines for astronomical hazard investigations.

• "Data-centrism" is important. Survey systems come in two main flavors. There are data-centric ones: those that put a rich and detailed specification of the data items/objects at the core of their architecture and at the "fingertips" of the user of the system. Such systems are in a sense the system analog of object oriented programming. There are the process-centric ones: those that put a rich and detailed specification of the data processing at the core of their architecture and at the "fingertips" of the user of the system. They are the system analog of functional programming. For the re-use of data items/objects for other purposes (like astronomical hazards), the rich and detailed specification of data items at the fingertips of the user make data re-use convenient. In the case of processing-centric architectures re-use of data becomes cumbersome if the dominant re-use is not in the re-use of processes.

5. Acknowledgments

This work was executed as part of ESA contract no. 4000134667/21/D/MRP (CARMEN) with their Planetary Defence Office. The pilots made use of the Big Data Layer of the Target Field Lab project "Mining Big Data". The Target Field Lab is supported by the Northern Netherlands Alliance (SNN) and is financially supported by the European Regional Development Fund. The data science software system AstroWISE runs on powerful databases and computing clusters at the Donald Smits Center of the University of Groningen and is supported, among other parties, by NOVA (the Dutch Research School for Astronomy). This research has made use of Aladin sky atlas (Bonnarel et al. 2000; Boch and Fernique 2014) developed at CDS, Strasbourg Observatory, France and SAOImageDS9 (Joye and Mandel 2003). This work has been done using the following software, packages and PYTHON libraries: Astro-WISE (Begeman et al. 2013, McFarland et al. 2013a), NUMPY (van der Walt et al. 2011), SCIPY (Virtanen et al. 2020), ASTROPY (Astropy Collaboration et al. 2018).

References

Abbott, T. M. C., Adamów, M., Aguena, M., Allam, S., Amon, A., Annis, J., Avila, S., Bacon, D., Banerji, M., Bechtol, K., Becker, M. R., Bernstein, G. M., Bertin, E., Bhargava, S., Bridle, S. L., Brooks, D., Burke, D. L., Carnero Rosell, A., Carrasco Kind, M., Carretero, J., Castander, F. J., Cawthon, R., Chang, C., Choi, A., Conselice, C., Costanzi, M., Crocce, M., da Costa, L. N., Davis, T. M., De Vicente, J., DeRose, J., Desai, S., Diehl, H. T., Dietrich, J. P., Drlica-Wagner, A., Eckert, K., Elvin-Poole, J., Everett, S., Evrard, A. E., Ferrero, I., Ferté, A., Flaugher, B., Fosalba, P., Friedel, D., Frieman, J., García-Bellido, J., Gaztanaga, E., Gelman, L., Gerdes, D. W., Giannantonio, T., Gill, M. S. S., Gruen, D., Gruendl, R. A., Gschwend, J., Gutierrez, G., Hartley, W. G., Hinton, S. R., Hollowood, D. L., Honscheid, K., Huterer, D., James, D. J., Jeltema, T., Johnson, M. D., Kent, S., Kron, R., Kuehn, K., Kuropatkin, N., Lahav, O., Li, T. S., Lidman, C., Lin, H., MacCrann, N., Maia, M. A. G., Manning, T. A., Maloney, J. D., March, M., Marshall, J. L., Martini, P., Melchior, P., Menanteau, F., Miquel, R., Morgan, R., Myles, J., Neilsen, E., Ogando, R. L. C., Palmese, A., Paz-Chinchón, F., Petravick, D., Pieres, A., Plazas, A. A., Pond, C., Rodriguez-Monroy, M., Romer, A. K., Roodman, A., Rykoff, E. S., Sako, M., Sanchez, E., Santiago, B., Scarpine, V., Serrano, S., Sevilla-Noarbe, I., Smith, J. A., Smith, M., Soares-Santos, M., Suchyta, E., Swanson, M. E. C., Tarle, G., Thomas, D., To, C., Tremblay, P. E., Troxel, M. A., Tucker, D. L., Turner, D. J., Varga, T. N., Walker, A. R., Wechsler, R. H., Weller, J., Wester, W., Wilkinson, R. D., Yanny, B., Zhang, Y., Nikutta, R., Fitzpatrick, M., Jacques, A., Scott, A., Olsen, K., Huang, L., Herrera, D., Juneau, S., Nidever, D., Weaver, B. A., Adean, C., Correia, V., de Freitas, M., Freitas, F. N., Singulani, C., Vila-Verde, G., & Linea Science Server 2021, The Dark Energy Survey Data Release 2. *ApJS*, 255(2), 20.
Astropy Collaboration, Price-Whelan, A. M., Sipőcz, B. M., Günther, H. M., Lim, P. L., Crawford, S. M., Conseil, S., Shupe, D. L., Craig, M. W., Dencheva, N., Ginsburg, A., Vand erPlas, J. T., Bradley, L. D., Pérez-Suárez, D., de Val-Borro, M., Aldcroft, T. L., Cruz, K. L., Robitaille, T. P., Tollerud, E. J., Ardelean, C., Babej, T., Bach, Y. P., Bachetti, M., Bakanov, A. V., Bamford, S. P., Barentsen, G., Barmby, P., Baumbach, A., Berry, K. L., Biscani, F., Boquien, M., Bostroem, K. A., Bouma, L. G., Brammer, G. B., Bray, E. M., Breytenbach, H., Buddelmeijer, H., Burke, D. J., Calderone, G., Cano Rodríguez, J. L., Cara, M., Cardoso, J. V. M., Cheedella, S., Copin, Y., Corrales, L., Crichton, D., D'Avella, D., Deil, C., Depagne, É., Dietrich, J. P., Donath, A., Droettboom, M., Earl, N., Erben, T., Fabbro, S., Ferreira, L. A., Finethy, T., Fox, R. T., Garrison, L. H., Gibbons, S. L. J.,

Goldstein, D. A., Gommers, R., Greco, J. P., Greenfield, P., Groener, A. M., Grollier, F., Hagen, A., Hirst, P., Homeier, D., Horton, A. J., Hosseinzadeh, G., Hu, L., Hunkeler, J. S., Ivezić, Ž., Jain, A., Jenness, T., Kanarek, G., Kendrew, S., Kern, N. S., Kerzendorf, W. E., Khvalko, A., King, J., Kirkby, D., Kulkarni, A. M., Kumar, A., Lee, A., Lenz, D., Littlefair, S. P., Ma, Z., Macleod, D. M., Mastropietro, M., McCully, C., Montagnac, S., Morris, B. M., Mueller, M., Mumford, S. J., Muna, D., Murphy, N. A., Nelson, S., Nguyen, G. H., Ninan, J. P., Nöthe, M., Ogaz, S., Oh, S., Parejko, J. K., Parley, N., Pascual, S., Patil, R., Patil, A. A., Plunkett, A. L., Prochaska, J. X., Rastogi, T., Reddy Janga, V., Sabater, J., Sakurikar, P., Seifert, M., Sherbert, L. E., Sherwood-Taylor, H., Shih, A. Y., Sick, J., Silbiger, M. T., Singanamalla, S., Singer, L. P., Sladen, P. H., Sooley, K. A., Sornarajah, S., Streicher, O., Teuben, P., Thomas, S. W., Tremblay, G. R., Turner, J. E. H., Terrón, V., van Kerkwijk, M. H., de la Vega, A., Watkins, L. L., Weaver, B. A., Whitmore, J. B., Woillez, J., Zabalza, V., & Astropy Contributors 2018, The Astropy Project: Building an Open-science Project and Status of the v2.0 Core Package. *AJ*, 156(3), 123.

Bailer-Jones, C. A. L. 2022, Stars That Approach within One Parsec of the Sun: New and More Accurate Encounters Identified in Gaia Data Release 3. *ApJL*, 935(1), L9.

Bailer-Jones, C. A. L., Rybizki, J., Andrae, R., & Fouesneau, M. 2018, New stellar encounters discovered in the second Gaia data release. *A&A*, 616, A37.

Begeman, K., Belikov, A. N., Boxhoorn, D. R., & Valentijn, E. A. 2013, The Astro-WISE datacentric information system. *Experimental Astronomy*, 35(1-2), 1–23.

Berthier, J., Carry, B., Mahlke, M., & Normand, J. 2022, SsODNet: The Solar system Open Database Network. *arXiv e-prints*,, arXiv:2209.10697.

Boch, T. & Fernique, P. Aladin Lite: Embed your Sky in the Browser. In Manset, N. & Forshay, P., editors, *Astronomical Data Analysis Software and Systems XXIII* 2014,, volume 485 of *Astronomical Society of the Pacific Conference Series*, 277.

Bonnarel, F., Fernique, P., Bienaymé, O., Egret, D., Genova, F., Louys, M., Ochsenbein, F., Wenger, M., & Bartlett, J. G. 2000, The ALADIN interactive sky atlas. A reference tool for identification of astronomical sources. *A&A Supplements*, 143, 33–40.

Borovička, J., Spurný, P., Brown, P., Wiegert, P., Kalenda, P., Clark, D., & Shrbený, L. 2013, The trajectory, structure and origin of the Chelyabinsk asteroidal impactor. *Nature*, 503(7475), 235–237.

Brown, P. G., Assink, J. D., Astiz, L., Blaauw, R., Boslough, M. B., Borovička, J., Brachet, N., Brown, D., Campbell-Brown, M., Ceranna, L., Cooke, W., de Groot-Hedlin, C., Drob, D. P., Edwards, W., Evers, L. G., Garces, M., Gill, J., Hedlin, M., Kingery, A., Laske, G., Le Pichon, A., Mialle, P., Moser, D. E., Saffer, A., Silber, E., Smets, P., Spalding, R. E., Spurný, P., Tagliaferri, E., Uren, D., Weryk, R. J., Whitaker, R., & Krzeminski, Z. 2013, A 500-kiloton airburst over Chelyabinsk and an enhanced hazard from small impactors. *Nature*, 503(7475), 238–241.

Brugger, J., Feulner, G., & Petri, S. 2017, Baby, it's cold outside: Climate model simulations of the effects of the asteroid impact at the end of the Cretaceous. *Geophysical Research Letters*, 44(1), 419–427.

Euclid Collaboration, Scaramella, R., Amiaux, J., Mellier, Y., Burigana, C., Carvalho, C. S., Cuillandre, J. C., Da Silva, A., Derosa, A., Dinis, J., Maiorano, E., Maris, M., Tereno, I., Laureijs, R., Boenke, T., Buenadicha, G., Dupac, X., Gaspar Venancio, L. M., Gómez-Álvarez, P., Hoar, J., Lorenzo Alvarez, J., Racca, G. D., Saavedra-Criado, G., Schwartz, J., Vavrek, R., Schirmer, M., Aussel, H., Azzollini, R., Cardone, V. F., Cropper, M., Ealet, A., Garilli, B., Gillard, W., Granett, B. R., Guzzo, L., Hoekstra, H., Jahnke, K., Kitching, T., Maciaszek, T., Meneghetti, M., Miller, L., Nakajima, R., Niemi, S. M., Pasian, F., Percival, W. J., Pottinger, S., Sauvage, M., Scodeggio, M., Wachter, S., Zacchei, A., Aghanim, N., Amara, A., Auphan, T., Auricchio, N., Awan, S., Balestra, A., Bender, R., Bodendorf, C., Bonino, D., Branchini, E., Brau-Nogue, S., Brescia, M., Candini, G. P., Capobianco, V., Carbone, C., Carlberg, R. G., Carretero, J., Casas, R., Castander, F. J., Castellano, M., Cavuoti, S., Cimatti, A., Cledassou, R., Congedo, G., Conselice, C. J., Conversi, L., Copin, Y., Corcione, L., Costille, A., Courbin, F., Degaudenzi, H., Douspis, M., Dubath,

F., Duncan, C. A. J., Dusini, S., Farrens, S., Ferriol, S., Fosalba, P., Fourmanoit, N., Frailis, M., Franceschi, E., Franzetti, P., Fumana, M., Gillis, B., Giocoli, C., Grazian, A., Grupp, F., Haugan, S. V. H., Holmes, W., Hormuth, F., Hudelot, P., Kermiche, S., Kiessling, A., Kilbinger, M., Kohley, R., Kubik, B., Kümmel, M., Kunz, M., Kurki-Suonio, H., Lahav, O., Ligori, S., Lilje, P. B., Lloro, I., Mansutti, O., Marggraf, O., Markovic, K., Marulli, F., Massey, R., Maurogordato, S., Melchior, M., Merlin, E., Meylan, G., Mohr, J. J., Moresco, M., Morin, B., Moscardini, L., Munari, E., Nichol, R. C., Padilla, C., Paltani, S., Peacock, J., Pedersen, K., Pettorino, V., Pires, S., Poncet, M., Popa, L., Pozzetti, L., Raison, F., Rebolo, R., Rhodes, J., Rix, H. W., Roncarelli, M., Rossetti, E., Saglia, R., Schneider, P., Schrabback, T., Secroun, A., Seidel, G., Serrano, S., Sirignano, C., Sirri, G., Skottfelt, J., Stanco, L., Starck, J. L., Tallada-Crespí, P., Tavagnacco, D., Taylor, A. N., Teplitz, H. I., Toledo-Moreo, R., Torradeflot, F., Trifoglio, M., Valentijn, E. A., Valenziano, L., Verdoes Kleijn, G. A., Wang, Y., Welikala, N., Weller, J., Wetzstein, M., Zamorani, G., Zoubian, J., Andreon, S., Baldi, M., Bardelli, S., Boucaud, A., Camera, S., Di Ferdinando, D., Fabbian, G., Farinelli, R., Galeotta, S., Graciá-Carpio, J., Maino, D., Medinaceli, E., Mei, S., Neissner, C., Polenta, G., Renzi, A., Romelli, E., Rosset, C., Sureau, F., Tenti, M., Vassallo, T., Zucca, E., Baccigalupi, C., Balaguera-Antolínez, A., Battaglia, P., Biviano, A., Borgani, S., Bozzo, E., Cabanac, R., Cappi, A., Casas, S., Castignani, G., Colodro-Conde, C., Coupon, J., Courtois, H. M., Cuby, J., de la Torre, S., Desai, S., Dole, H., Fabricius, M., Farina, M., Ferreira, P. G., Finelli, F., Flose-Reimberg, P., Fotopoulou, S., Ganga, K., Gozaliasl, G., Hook, I. M., Keihanen, E., Kirkpatrick, C. C., Liebing, P., Lindholm, V., Mainetti, G., Martinelli, M., Martinet, N., Maturi, M., McCracken, H. J., Metcalf, R. B., Morgante, G., Nightingale, J., Nucita, A., Patrizii, L., Potter, D., Riccio, G., Sánchez, A. G., Sapone, D., Schewtschenko, J. A., Schultheis, M., Scottez, V., Teyssier, R., Tutusaus, I., Valiviita, J., Viel, M., Vriend, W., & Whittaker, L. 2022, Euclid preparation. I. The Euclid Wide Survey. *A&A*, 662, A112.

Gaia Collaboration, Vallenari, A., Brown, A. G. A., Prusti, T., de Bruijne, J. H. J., Arenou, F., Babusiaux, C., Biermann, M., Creevey, O. L., Ducourant, C., Evans, D. W., Eyer, L., Guerra, R., Hutton, A., Jordi, C., Klioner, S. A., Lammers, U. L., Lindegren, L., Luri, X., Mignard, F., Panem, C., Pourbaix, D., Randich, S., Sartoretti, P., Soubiran, C., Tanga, P., Walton, N. A., Bailer-Jones, C. A. L., Bastian, U., Drimmel, R., Jansen, F., Katz, D., Lattanzi, M. G., van Leeuwen, F., Bakker, J., Cacciari, C., Castañeda, J., De Angeli, F., Fabricius, C., Fouesneau, M., Frémat, Y., Galluccio, L., Guerrier, A., Heiter, U., Masana, E., Messineo, R., Mowlavi, N., Nicolas, C., Nienartowicz, K., Pailler, F., Panuzzo, P., Riclet, F., Roux, W., Seabroke, G. M., Sordoørcit, R., Thévenin, F., Gracia-Abril, G., Portell, J., Teyssier, D., Altmann, M., Andrae, R., Audard, M., Bellas-Velidis, I., Benson, K., Berthier, J., Blomme, R., Burgess, P. W., Busonero, D., Busso, G., Cánovas, H., Carry, B., Cellino, A., Cheek, N., Clementini, G., Damerdji, Y., Davidson, M., de Teodoro, P., Nuñez Campos, M., Delchambre, L., Dell'Oro, A., Esquej, P., Fernández-Hernández, J., Fraile, E., Garabato, D., García-Lario, P., Gosset, E., Haigron, R., Halbwachs, J. L., Hambly, N. C., Harrison, D. L., Hernández, J., Hestroffer, D., Hodgkin, S. T., Holl, B., Janßen, K., Jevardat de Fombelle, G., Jordan, S., Krone-Martins, A., Lanzafame, A. C., Löffler, W., Marchal, O., Marrese, P. M., Moitinho, A., Muinonen, K., Osborne, P., Pancino, E., Pauwels, T., Recio-Blanco, A., Reylé, C., Riello, M., Rimoldini, L., Roegiers, T., Rybizki, J., Sarro, L. M., Siopis, C., Smith, M., Sozzetti, A., Utrilla, E., van Leeuwen, M., Abbas, U., Ábrahám, P., Abreu Aramburu, A., Aerts, C., Aguado, J. J., Ajaj, M., Aldea-Montero, F., Altavilla, G., Álvarez, M. A., Alves, J., Anders, F., Anderson, R. I., Anglada Varela, E., Antoja, T., Baines, D., Baker, S. G., Balaguer-Núñez, L., Balbinot, E., Balog, Z., Barache, C., Barbato, D., Barros, M., Barstow, M. A., Bartolomé, S., Bassilana, J. L., Bauchet, N., Becciani, U., Bellazzini, M., Berihuete, A., Bernet, M., Bertone, S., Bianchi, L., Binnenfeld, A., Blanco-Cuaresma, S., Blazere, A., Boch, T., Bombrun, A., Bossini, D., Bouquillon, S., Bragaglia, A., Bramante, L., Breedt, E., Bressan, A., Brouillet, N., Brugaletta, E., Bucciarelli, B., Burlacu, A., Butkevich, A. G., Buzzi, R., Caffau, E., Cancelliere, R., Cantat-Gaudin, T., Carballo, R., Carlucci, T., Carnerero, M. I., Carrasco, J. M., Casamiquela,

L., Castellani, M., Castro-Ginard, A., Chaoul, L., Charlot, P., Chemin, L., Chiaramida, V., Chiavassa, A., Chornay, N., Comoretto, G., Contursi, G., Cooper, W. J., Cornez, T., Cowell, S., Crifo, F., Cropper, M., Crosta, M., Crowley, C., Dafonte, C., Dapergolas, A., David, M., David, P., de Laverny, P., De Luise, F., De March, R., De Ridder, J., de Souza, R., de Torres, A., del Peloso, E. F., del Pozo, E., Delbo, M., Delgado, A., Delisle, J. B., Demouchy, C., Dharmawardena, T. E., Di Matteo, P., Diakite, S., Diener, C., Distefano, E., Dolding, C., Edvardsson, B., Enke, H., Fabre, C., Fabrizio, M., Faigler, S., Fedorets, G., Fernique, P., Fienga, A., Figueras, F., Fournier, Y., Fouron, C., Fragkoudi, F., Gai, M., Garcia-Gutierrez, A., Garcia-Reinaldos, M., García-Torres, M., Garofalo, A., Gavel, A., Gavras, P., Gerlach, E., Geyer, R., Giacobbe, P., Gilmore, G., Girona, S., Giuffrida, G., Gomel, R., Gomez, A., González-Núñez, J., González-Santamaría, I., González-Vidal, J. J., Granvik, M., Guillout, P., Guiraud, J., Gutiérrez-Sánchez, R., Guy, L. P., Hatzidimitriou, D., Hauser, M., Haywood, M., Helmer, A., Helmi, A., Sarmiento, M. H., Hidalgo, S. L., Hilger, T., Hładczuk, N., Hobbs, D., Holland, G., Huckle, H. E., Jardine, K., Jasniewicz, G., Jean-Antoine Piccolo, A., Jiménez-Arranz, Ó., Jorissen, A., Juaristi Campillo, J., Julbe, F., Karbevska, L., Kervella, P., Khanna, S., Kontizas, M., Kordopatis, G., Korn, A. J., Kóspál, Á., Kostrzewa-Rutkowska, Z., Kruszyńska, K., Kun, M., Laizeau, P., Lambert, S., Lanza, A. F., Lasne, Y., Le Campion, J. F., Lebreton, Y., Lebzelter, T., Leccia, S., Leclerc, N., Lecoeur-Taibi, I., Liao, S., Licata, E. L., Lindstrøm, H. E. P., Lister, T. A., Livanou, E., Lobel, A., Lorca, A. 2022, Gaia Data Release 3: Summary of the content and survey properties. *arXiv e-prints,*, arXiv:2208.00211.

Gwyn, S. D. J., Hill, N., & Kavelaars, J. J. 2012, SSOS: A Moving-Object Image Search Tool for Asteroid Precovery. *PASP*, 124(916), 579.

Ibata, R. A., McConnachie, A., Cuillandre, J.-C., Fantin, N., Haywood, M., Martin, N. F., Bergeron, P., Beckmann, V., Bernard, E., Bonifacio, P., Caffau, E., Carlberg, R., Côté, P., Cabanac, R., Chapman, S., Duc, P.-A., Durret, F., Famaey, B., Fabbro, S., Gwyn, S., Hammer, F., Hill, V., Hudson, M. J., Lançon, A., Lewis, G., Malhan, K., di Matteo, P., McCracken, H., Mei, S., Mellier, Y., Navarro, J., Pires, S., Pritchet, C., Reylé, C., Richer, H., Robin, A. C., Sánchez-Janssen, R., Sawicki, M., Scott, D., Scottez, V., Spekkens, K., Starkenburg, E., Thomas, G., & Venn, K. 2017, The Canada-France Imaging Survey: First Results from the u-Band Component. *ApJ*, 848(2), 128.

Ivezić, Ž., Kahn, S. M., Tyson, J. A., Abel, B., Acosta, E., Allsman, R., Alonso, D., AlSayyad, Y., Anderson, S. F., Andrew, J., Angel, J. R. P., Angeli, G. Z., Ansari, R., Antilogus, P., Araujo, C., Armstrong, R., Arndt, K. T., Astier, P., Aubourg, É., Auza, N., Axelrod, T. S., Bard, D. J., Barr, J. D., Barrau, A., Bartlett, J. G., Bauer, A. E., Bauman, B. J., Baumont, S., Bechtol, E., Bechtol, K., Becker, A. C., Becla, J., Beldica, C., Bellavia, S., Bianco, F. B., Biswas, R., Blanc, G., Blazek, J., Blandford, R. D., Bloom, J. S., Bogart, J., Bond, T. W., Booth, M. T., Borgland, A. W., Borne, K., Bosch, J. F., Boutigny, D., Brackett, C. A., Bradshaw, A., Brandt, W. N., Brown, M. E., Bullock, J. S., Burchat, P., Burke, D. L., Cagnoli, G., Calabrese, D., Callahan, S., Callen, A. L., Carlin, J. L., Carlson, E. L., Chandrasekharan, S., Charles-Emerson, G., Chesley, S., Cheu, E. C., Chiang, H.-F., Chiang, J., Chirino, C., Chow, D., Ciardi, D. R., Claver, C. F., Cohen-Tanugi, J., Cockrum, J. J., Coles, R., Connolly, A. J., Cook, K. H., Cooray, A., Covey, K. R., Cribbs, C., Cui, W., Cutri, R., Daly, P. N., Daniel, S. F., Daruich, F., Daubard, G., Daues, G., Dawson, W., Delgado, F., Dellapenna, A., de Peyster, R., de Val-Borro, M., Digel, S. W., Doherty, P., Dubois, R., Dubois-Felsmann, G. P., Durech, J., Economou, F., Eifler, T., Eracleous, M., Emmons, B. L., Fausti Neto, A., Ferguson, H., Figueroa, E., Fisher-Levine, M., Focke, W., Foss, M. D., Frank, J., Freemon, M. D., Gangler, E., Gawiser, E., Geary, J. C., Gee, P., Geha, M., Gessner, C. J. B., Gibson, R. R., Gilmore, D. K., Glanzman, T., Glick, W., Goldina, T., Goldstein, D. A., Goodenow, I., Graham, M. L., Gressler, W. J., Gris, P., Guy, L. P., Guyonnet, A., Haller, G., Harris, R., Hascall, P. A., Haupt, J., Hernandez, F., Herrmann, S., Hileman, E., Hoblitt, J., Hodgson, J. A., Hogan, C., Howard, J. D., Huang, D., Huffer, M. E., Ingraham, P., Innes, W. R., Jacoby, S. H., Jain, B., Jammes, F., Jee, M. J., Jenness, T., Jernigan, G., Jevremović, D., Johns, K., Johnson, A. S., Johnson,

M. W. G., Jones, R. L., Juramy-Gilles, C., Jurić, M., Kalirai, J. S., Kallivayalil, N. J., Kalmbach, B., Kantor, J. P., Karst, P., Kasliwal, M. M., Kelly, H., Kessler, R., Kinnison, V., Kirkby, D., Knox, L., Kotov, I. V., Krabbendam, V. L., Krughoff, K. S., Kubánek, P., Kuczewski, J., Kulkarni, S., Ku, J., Kurita, N. R., Lage, C. S., Lambert, R., Lange, T., Langton, J. B., Le Guillou, L., Levine, D., Liang, M., Lim, K.-T., Lintott, C. J., Long, K. E., Lopez, M., Lotz, P. J., Lupton, R. H., Lust, N. B., MacArthur, L. A., Mahabal, A., Mandelbaum, R., Markiewicz, T. W., Marsh, D. S., Marshall, P. J., Marshall, S., May, M., McKercher, R., McQueen, M., Meyers, J., Migliore, M., Miller, M., Mills, D. J., Miraval, C., Moeyens, J., Moolekamp, F. E., Monet, D. G., Moniez, M., Monkewitz, S., Montgomery, C., Morrison, C. B., Mueller, F., Muller, G. P., Muñoz Arancibia, F., Neill, D. R., Newbry, S. P., Nief, J.-Y., Nomerotski, A., Nordby, M., O'Connor, P., Oliver, J., Olivier, S. S., Olsen, K., O'Mullane, W., Ortiz, S., Osier, S., Owen, R. E., Pain, R., Palecek, P. E., Parejko, J. K., Parsons, J. B., Pease, N. M., Peterson, J. M., Peterson, J. R., Petravick, D. L., Libby Petrick, M. E., Petry, C. E., Pierfederici, F., Pietrowicz, S., Pike, R., Pinto, P. A., Plante, R., Plate, S., Plutchak, J. P., Price, P. A., Prouza, M., Radeka, V., Rajagopal, J., Rasmussen, A. P., Regnault, N., Reil, K. A., Reiss, D. J., Reuter, M. A., Ridgway, S. T., Riot, V. J., Ritz, S., Robinson, S., Roby, W., Roodman, A., Rosing, W., Roucelle, C., Rumore, M. R., Russo, S., Saha, A., Sassolas, B., Schalk, T. L., Schellart, P., Schindler, R. H., Schmidt, S., Schneider, D. P., Schneider, M. D., Schoening, W., Schumacher, G., Schwamb, M. E., Sebag, J., Selvy, B., Sembroski, G. H., Seppala, L. G., Serio, A., Serrano, E., Shaw, R. A., Shipsey, I., Sick, J., Silvestri, N., Slater, C. T., Smith, J. A., Smith, R. C., Sobhani, S., Soldahl, C., Storrie-Lombardi, L., Stover, E., Strauss, M. A., Street, R. A., Stubbs, C. W., Sullivan, I. S., Sweeney, D., Swinbank, J. D., Szalay, A., Takacs, P., Tether, S. A., Thaler, J. J., Thayer, J. G., Thomas, S., Thornton, A. J., Thukral, V., Tice, J., Trilling, D. E., Turri, M., Van Berg, R., Vanden Berk, D., Vetter, K., Virieux, F., Vucina, T., Wahl, W 2019, LSST: From Science Drivers to Reference Design and Anticipated Data Products. *ApJ*, 873(2), 111.

Joye, W. A. & Mandel, E. New Features of SAOImage DS9. In Payne, H. E., Jedrzejewski, R. I., & Hook, R. N., editors, *Astronomical Data Analysis Software and Systems XII* 2003,, volume 295 of *Astronomical Society of the Pacific Conference Series*, 489.

Kaiser, N., Burgett, W., Chambers, K., Denneau, L., Heasley, J., Jedicke, R., Magnier, E., Morgan, J., Onaka, P., & Tonry, J. The Pan-STARRS wide-field optical/NIR imaging survey. In Stepp, L. M., Gilmozzi, R., & Hall, H. J., editors, *Ground-based and Airborne Telescopes III* 2010,, volume 7733 of *Society of Photo-Optical Instrumentation Engineers (SPIE) Conference Series*, 77330E.

Kuijken, K. *et al.* 2019, The fourth data release of the Kilo-Degree Survey: ugri imaging and nine-band optical-IR photometry over 1000 square degrees. *A&A*, 625, A2.

Laureijs, R., Amiaux, J., Arduini, S., Auguères, J. L., Brinchmann, J., Cole, R., Cropper, M., Dabin, C., Duvet, L., Ealet, A., Garilli, B., Gondoin, P., Guzzo, L., Hoar, J., Hoekstra, H., Holmes, R., Kitching, T., Maciaszek, T., Mellier, Y., Pasian, F., Percival, W., Rhodes, J., Saavedra Criado, G., Sauvage, M., Scaramella, R., Valenziano, L., Warren, S., Bender, R., Castander, F., Cimatti, A., Le Fèvre, O., Kurki-Suonio, H., Levi, M., Lilje, P., Meylan, G., Nichol, R., Pedersen, K., Popa, V., Rebolo Lopez, R., Rix, H. W., Rottgering, H., Zeilinger, W., Grupp, F., Hudelot, P., Massey, R., Meneghetti, M., Miller, L., Paltani, S., Paulin-Henriksson, S., Pires, S., Saxton, C., Schrabback, T., Seidel, G., Walsh, J., Aghanim, N., Amendola, L., Bartlett, J., Baccigalupi, C., Beaulieu, J. P., Benabed, K., Cuby, J. G., Elbaz, D., Fosalba, P., Gavazzi, G., Helmi, A., Hook, I., Irwin, M., Kneib, J. P., Kunz, M., Mannucci, F., Moscardini, L., Tao, C., Teyssier, R., Weller, J., Zamorani, G., Zapatero Osorio, M. R., Boulade, O., Foumond, J. J., Di Giorgio, A., Guttridge, P., James, A., Kemp, M., Martignac, J., Spencer, A., Walton, D., Blümchen, T., Bonoli, C., Bortoletto, F., Cerna, C., Corcione, L., Fabron, C., Jahnke, K., Ligori, S., Madrid, F., Martin, L., Morgante, G., Pamplona, T., Prieto, E., Riva, M., Toledo, R., Trifoglio, M., Zerbi, F., Abdalla, F., Douspis, M., Grenet, C., Borgani, S., Bouwens, R., Courbin, F., Delouis, J. M., Dubath, P., Fontana, A., Frailis, M., Grazian, A., Koppenhöfer, J., Mansutti, O.,

Melchior, M., Mignoli, M., Mohr, J., Neissner, C., Noddle, K., Poncet, M., Scodeggio, M., Serrano, S., Shane, N., Starck, J. L., Surace, C., Taylor, A., Verdoes-Kleijn, G., Vuerli, C., Williams, O. R., Zacchei, A., Altieri, B., Escudero Sanz, I., Kohley, R., Oosterbroek, T., Astier, P., Bacon, D., Bardelli, S., Baugh, C., Bellagamba, F., Benoist, C., Bianchi, D., Biviano, A., Branchini, E., Carbone, C., Cardone, V., Clements, D., Colombi, S., Conselice, C., Cresci, G., Deacon, N., Dunlop, J., Fedeli, C., Fontanot, F., Franzetti, P., Giocoli, C., Garcia-Bellido, J., Gow, J., Heavens, A., Hewett, P., Heymans, C., Holland, A., Huang, Z., Ilbert, O., Joachimi, B., Jennins, E., Kerins, E., Kiessling, A., Kirk, D., Kotak, R., Krause, O., Lahav, O., van Leeuwen, F., Lesgourgues, J., Lombardi, M., Magliocchetti, M., Maguire, K., Majerotto, E., Maoli, R., Marulli, F., Maurogordato, S., McCracken, H., McLure, R., Melchiorri, A., Merson, A., Moresco, M., Nonino, M., Norberg, P., Peacock, J., Pello, R., Penny, M., Pettorino, V., Di Porto, C., Pozzetti, L., Quercellini, C., Radovich, M., Rassat, A., Roche, N., Ronayette, S., Rossetti, E., Sartoris, B., Schneider, P., Semboloni, E., Serjeant, S., Simpson, F., Skordis, C., Smadja, G., Smartt, S., Spano, P., Spiro, S., Sullivan, M., Tilquin, A., Trotta, R., Verde, L., Wang, Y., Williger, G., Zhao, G., Zoubian, J., & Zucca, E. 2011, Euclid Definition Study Report. *arXiv e-prints,*, arXiv:1110.3193.

McFarland, J. P., Verdoes-Kleijn, G., Sikkema, G., Helmich, E. M., Boxhoorn, D. R., & Valentijn, E. A. 2013,a The Astro-WISE optical image pipeline. Development and implementation. *Experimental Astronomy*, 35a(1-2), 45–78.

McFarland, J. P., Verdoes-Kleijn, G., Sikkema, G., Helmich, E. M., Boxhoorn, D. R., & Valentijn, E. A. 2013,b The Astro-WISE optical image pipeline. Development and implementation. *Experimental Astronomy*, 35b(1-2), 45–78.

Mulder, W., de Jong, J. T. A., Verdoes Kleijn, G. A., Valentijn, E. A., Williams, O. R., Boxhoorn, D. R., Belikov, A. N., Fabricius, M., Helmich, E. M., Wetzstein, M., Vassallo, T., & George, K. The Astrometric Calibration Software System for Euclid's External Surveys. In Pizzo, R., Deul, E. R., Mol, J. D., de Plaa, J., & Verkouter, H., editors, *Astronomical Data Analysis Software and Systems XXIX* 2020,, volume 527 of *Astronomical Society of the Pacific Conference Series*, 615.

Nieto, S., de Teodoro, P., Salgado, J., Altieri, B., Buenadicha, G., Belikov, A., Boxhoorn, D., McFarland, J., Valentijn, E. A., Williams, O. R., Droege, B., & Tsyganov, A. The Euclid Archive System: A Data-Centric Approach to Big Data. In Molinaro, M., Shortridge, K., & Pasian, F., editors, *Astronomical Data Analysis Software and Systems XXVI* 2019,, volume 521 of *Astronomical Society of the Pacific Conference Series,* 12.

Peletier, R., Iodice, E., Venhola, A., Capaccioli, M., Cantiello, M., D'Abrusco, R., Falcón-Barroso, J., Grado, A., Hilker, M., Limatola, L., Mieske, S., Napolitano, N., Paolillo, M., Spavone, M., Valentijn, E., van de Ven, G., & Verdoes Kleijn, G. 2020, The Fornax Deep Survey data release 1. *arXiv e-prints,*, arXiv:2008.12633.

Pöntinen, M., Granvik, M., Nucita, A., Conversi, L., Altieri, B., *et al.* 2020, Euclid: Identification of asteroid streaks in simulated images using streakdet software. *Astronomy & Astrophysics*, 644, A35.

Pöntinen, M., Granvik, M., Nucita, A. A., Conversi, L., Altieri, B., Auricchio, N., Bodendorf, C., Bonino, D., Brescia, M., Capobianco, V., Carretero, J., Carry, B., Castellano, M., Cledassou, R., Congedo, G., Corcione, L., Cropper, M., Dusini, S., Frailis, M., Franceschi, E., Fumana, M., Garilli, B., Grupp, F., Hormuth, F., Israel, H., Jahnke, K., Kermiche, S., Kitching, T., Kohley, R., Kubik, B., Kunz, M., Laureijs, R., Lilje, P. B., Lloro, I., Maiorano, E., Marggraf, O., Massey, R., Meneghetti, M., Meylan, G., Moscardini, L., Padilla, C., Paltani, S., Pasian, F., Pires, S., Polenta, G., Raison, F., Roncarelli, M., Rossetti, E., Saglia, R., Schneider, P., Secroun, A., Serrano, S., Sirri, G., Taylor, A. N., Tereno, I., Toledo-Moreo, R., Valenziano, L., Wang, Y., Wetzstein, M., & Zoubian, J. 2020, Euclid: Identification of asteroid streaks in simulated images using StreakDet software. *A&A*, 644, A35.

Popova, O. P., Jenniskens, P., Emel'yanenko, V., Kartashova, A., Biryukov, E., Khaibrakhmanov, S., Shuvalov, V., Rybnov, Y., Dudorov, A., Grokhovsky, V. I., Badyukov, D. D., Yin, Q.-Z., Gural, P. S., Albers, J., Granvik, M., Evers, L. G., Kuiper, J., Kharlamov,

V., Solovyov, A., Rusakov, Y. S., Korotkiy, S., Serdyuk, I., Korochantsev, A. V., Larionov, M. Y., Glazachev, D., Mayer, A. E., Gisler, G., Gladkovsky, S. V., Wimpenny, J., Sanborn, M. E., Yamakawa, A., Verosub, K. L., Rowland, D. J., Roeske, S., Botto, N. W., Friedrich, J. M., Zolensky, M. E., Le, L., Ross, D., Ziegler, K., Nakamura, T., Ahn, I., Lee, J. I., Zhou, Q., Li, X.-H., Li, Q.-L., Liu, Y., Tang, G.-Q., Hiroi, T., Sears, D., Weinstein, I. A., Vokhmintsev, A. S., Ishchenko, A. V., Schmitt-Kopplin, P., Hertkorn, N., Nagao, K., Haba, M. K., Komatsu, M., Mikouchi, T., & aff34 2013, Chelyabinsk Airburst, Damage Assessment, Meteorite Recovery, and Characterization. *Science*, 342(6162), 1069–1073.

Portegies Zwart, S. 2021, Oort cloud Ecology. I. Extra-solar Oort clouds and the origin of asteroidal interlopers. *A&A*, 647, A136.

Rybizki, J., Demleitner, M., Fouesneau, M., Bailer-Jones, C., Rix, H.-W., & Andrae, R. 2018, A Gaia DR2 Mock Stellar Catalog. *PASP*, 130(989), 074101.

Saifollahi, T., V.-K. G. W. O. 2023, Mining archival data from wide-field astronomical surveys in search of near-Earth objects. *A&A*,.

Sharma, S., Bland-Hawthorn, J., Johnston, K. V., & Binney, J. 2011, Galaxia: A Code to Generate a Synthetic Survey of the Milky Way. *ApJ*, 730(1), 3.

Speagle, J. S. 2020,. *Mapping the Milky Way in the age of Gaia*. PhD thesis, Harvard University, Massachusetts.

Tian, H.-J., Gupta, P., Sesar, B., Rix, H.-W., Martin, N. F., Liu, C., Goldman, B., Platais, I., Kudritzki, R.-P., & Waters, C. Z. 2017, A Gaia-PS1-SDSS (GPS1) Proper Motion Catalog Covering 3/4 of the Sky. *ApJS*, 232(1), 4.

Tyson, J. A. Cosmology data analysis challenges and opportunities in the LSST sky survey. In *Journal of Physics Conference Series* 2019,, volume 1290 of *Journal of Physics Conference Series*, 012001.

van der Walt, S., Colbert, S. C., & Varoquaux, G. 2011, The NumPy Array: A Structure for Efficient Numerical Computation. *Computing in Science and Engineering*, 13(2), 22–30.

Verdoes Kleijn, G. A. e. a. Object classification with Convolutional Neural Networks: from KiDS to Euclid. In Gwyn, S. e. a., editor, *Astronomical Data Analysis Software and Systems XXXII* 2023,, volume 530 of *Astronomical Society of the Pacific Conference Series*, in press.

Virtanen, J., Poikonen, J., Säntti, T., Komulainen, T., Torppa, J., Granvik, M., Muinonen, K., Pentikäinen, H., Martikainen, J., Näränen, J., Lehti, J., & Flohrer, T. 2016, Streak detection and analysis pipeline for space-debris optical images. *Advances in Space Research*, 57(8), 1607–1623.

Virtanen, P., Gommers, R., Oliphant, T. E., Haberland, M., Reddy, T., Cournapeau, D., Burovski, E., Peterson, P., Weckesser, W., Bright, J., van der Walt, S. J., Brett, M., Wilson, J., Millman, K. J., Mayorov, N., Nelson, A. R. J., Jones, E., Kern, R., Larson, E., Carey, C. J., Polat, İ., Feng, Y., Moore, E. W., VanderPlas, J., Laxalde, D., Perktold, J., Cimrman, R., Henriksen, I., Quintero, E. A., Harris, C. R., Archibald, A. M., Ribeiro, A. H., Pedregosa, F., van Mulbregt, P., & SciPy 1.0 Contributors 2020, SciPy 1.0: Fundamental Algorithms for Scientific Computing in Python. *Nature Methods*, 17, 261–272.

Astronomical Hazards for Life on Earth
Proceedings IAU Symposium No. 374, 2025
G. Tancredi, ed.
doi:10.1017/S1743921324000759

Dynamic evolution scale of NEA population: dependence on the initial orbital parameters

Boris Shustov[ID] and Roman Zolotarev[ID]

Institute of Astronomy of the Russian Academy of Sciences,
Pyatnitskaya Street, 48, Moscow, 119017, Russia
email: bshustov@mail.ru, rv_zolotarev@mail.ru

Abstract. The history of crater formation on the Moon idicates that the number of NEAs larger than 50 m practically did not change over the past 2–3 Gyr. On the other hand a dynamic scale of the NEA population, which could be characterized by the depletion time by half t_{NEA}, is many orders of magnitude shorter. There are significant variations of t_{NEA} estimates by other authors. It is important to know this value more precisely, since this knowledge imposes restrictions on the mechanisms of replenishment of the NEAs, the lifetime of the Main Asteroid Belt, etc. In the Zolotarev & Shustov (2021) we have estimated t_{NEA} as 3.5 million years. We noted either that t_{NEA} for subgroups of NEAs depends on the initial orbital parameters of the subgroups. In the current study we considered this dependence quantitively. We have integrated orbits of 10 000 asteroids larger than 1 km and $q < 1.72$ AU over 20 Myr. These sample essentially includes all large NEAs (> 1 km). The NEA subsample is considered to be complete. We made integrations with the REBOUND software package using the MERCURIUS hybrid scheme (Rein et al. (2019)). To reveal dependence of t_{NEA} on orbital parameters $t_{NEA}(a, e, i)$ we divided the NEA subsample into 18 subgroups according to their orbital parameters. We found that t_{NEA} is substantially higher for subgroups with higher i and e. There is strong dependence of t_{NEA} on a. All these dependencies are explained by a different number of close approaches of asteroids from NEA subgroups to planets. We found that depletion of total NEO population can be approximated remarkably well with the following expression: $N(t)/N_0 = exp(-0.5 \times t^{0.33})$ where N_0 is an initial number and $N(t)$ – a current number of NEAs.

Keywords. asteroids, NEA

1. Introduction

Studies of dynamic properties of NEA popuation is of particular importance in the context of the problems of asteroid-comet hazard and asteroid resources. One of the most intriguing features of the evolution is that the number of NEAs remained almost unchanged over the past 2–3 billion years, although the dynamic scale of the existence of the current population is relatively short. The number of NEAs in previous epochs can be judged only by analyzing the change in the rate of impact crater formation on surfaces of non-atmospheric bodies of the Solar System. The Moon seems to be most convenient natural "log" for studying records of collisions over a long time interval (billions of years). The absence of atmosphere, water, and tectonic activity contributes to the preservation of this log. The history of the bombardment of the Moon can be described by the formula from Neukum & Ivanov (2002),

$$N(1) = 5.44 \cdot 10^{-14}(\exp(6.93t) 1) + 8.38 \cdot 10^{-4}t, \tag{1}$$

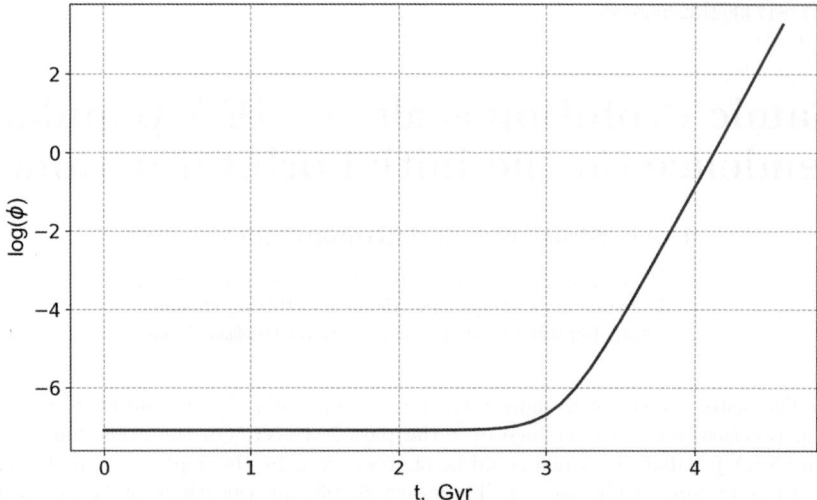

Figure 1. History of crater formation rate on the Moon. ϕ is a number of large (> 1 km) craters formed at 1 km^2 area per year. t is given in billion years from our epoch to the past.

where $N(1)$ is the number of craters larger than 1 km formed on the surface of the Moon per 1 km^2 during t, which is given in billion years from our epoch to the past. The time derivative of this function gives an estimate of the bombardment rate (see Fig. 1.)

From Fig. 1 several conclusions can be drawn. First, the rate of bombardment of the Moon as well as the Earth by bodies larger than 50 m (they are believed to be responsible for the formation of impact craters larger than 1 km) has changed little, although ~ 4 billion years ago the rate of bombardment was many orders of magnitude higher. The second, bodies $\geqslant 50$ m collide the Earth on average once every few hundred years. Of course, the estimates obtained above give an estimate of an average rate of bombardment. On scale of billions years various events could cause (temporary) changes in this rate. E.g. estimates of changes in the number of near-Earth objects based on the analysis of the age of lunar craters over the last billion years led Mazrouei et al. (2019); Ipatov et al. (2020) to conclude that the number of collisions per time unit increased 2.6 times about 290 million years ago. The possible cause of such fluctuations is destructive collisions of large bodies. Thus, according to Bottke et al. (2007), the catastrophic destruction of a large Main Belt asteroid occured about 160 million years ago could almost double the number of NEAs larger than 1 km compared to the average number for an interval of 1 billion years. One more reason of fluctuations is dynamic disturbances in the population of minor bodies caused by the Solar System's encounters to stars. In Shustov et al. (2020) it is shown that the Solar System approaches a star (at a distance of $\sim 10^5$ AU) several times (on average 4) over 1 million years. Closer passes of stars at a distance of $\sim 10^3$ AU that cause strong dynamic disturbances in the ensemble of minor bodies of the Solar System occur approximately once every 1–2 billion years.

Since NEAs are responsible for the formation of impact craters on the Moon (the contribution of comets is much smaller) it is logical to compare the statistics of lunar (planetary) data with astronomical data on the NEA size distribution. As noted in Strom et al. (2015), in general, lunar data on the statistics (sizes) of impactors and astronomical data on the NEA size distribution are in good agreement, but they differ significantly from the statistics of the sizes of asteroids of the Main Belt.

According to Morbidelli et al. (2002) the main source of NEAs ($\sim 95\%$) is the Main Belt (MB), the source of the remaining NEAs is the Kuiper Belt. This means that it is necessary to study the origin and properties of the NEAs together with the study of the entire ensemble of MB asteroids. The interplay between the two populations (NEAs and MB asteroids) include at least the following (obvious) questions:

• how many of the refugees could come back into the MB?

• what is the net asteroid loss rate from the MB (in other words how long MB will survive because of loss of NEAs)?

• what is the dynamic fate of the NEA population?

• etc.

So far, a comprehensive model which could give "all" the answers does not exist although in recent years very interesting models considering some of the issues have been appeared.

Here we focus on dynamic fate of the NEA population. Most remarkable feature is relatively short dynamic time scale (in compare to very long period of constant NEA flux entering Moon surface as it is illustrated in Fig. 1). The elongated orbits of NEAs are subject to strong disturbances from the planets and on this short time scale they either abandon the Solar System or fall onto the Sun and planets or they leave the scope of definition of NEA (NEA zone, where $q \leqslant 1.3$ AU). Unlike the previous two the latter process is not irreversible one. Some NEAs leave the NEA zone, and vice versa some asteroids from the non-NEA area enter NEA zone. For simplicity and by physical analogy we call this phenomenon diffusion in the space of parameters of asteroidal orbits or simply diffusion of orbits.

First estimates of the dynamic scale of the NEA population by Farinella et al. (1994), Gladman et al. (2000), O'Brien et al. (2003) demonstrated that this scale is of order of several million years. Usually the so-called median time, i.e. the time interval during which the population of NEAs is halved, is used to evaluate the scale. We denote this time scale as t_{NEA}. In Gladman et al. (2000), the dynamic evolution of a sample of 117 NEAs was followed over a time interval of 60 million years. It was shown that 10–20% NEAs from the sample end their lives by colliding Venus or Earth, more than half end by falling onto the Sun, and about 15% are ejected from the Solar System. The median lifetime for this sample was estimated $\sim$10 million years. In a more recent paper by Granvik et al. (2018) dynamical evolution of NEA ensemble was studied taking into account their sizes. The authors estimated the average NEO lifetime as $\lesssim 10$ million years. Note that definition of lifetime in Granvik et al. (2018) is not same as t_{NEA} defined above, it is a time spent in the parameter region ($q < 1.3$ AU and $a < 4.2$ AU). Dependence of lifetime on asteroid size is not very strong and ranges from about 6 to 11 Myr for asteroids of size $2 - 0.05$ km with the mid-sized NEOs having the shortest lifetimes.

We think that it is important to know t_{NEA} more precisely, since this knowledge imposes restrictions on the mechanisms of replenishment of the NEA population, on estimates of the number of NEAs, on the lifetime of MB, etc. In Zolotarev & Shustov (2021), we showed that the lifetime of the current NEA population is about 3 million years. We also noted that the t_{NEA} strongly depends on initial orbital parameters (a, e) of NEAs. In this paper the dependence on the initial parameters of asteroidal orbits is studied in more detail. In section 2 the task formulation and method of modelling are presented; section 3 is devoted to the results obtained. The conclusions are given in section 4.

2. Problem statement and calculation method

One of the tasks of this work was essentially to repeat the t_{NEA} estimate but using complete data set. Currently the number of known NEAs is $\sim$28 000. However this set is obviously incomplete. We have to consider evolution of a group of initially selected

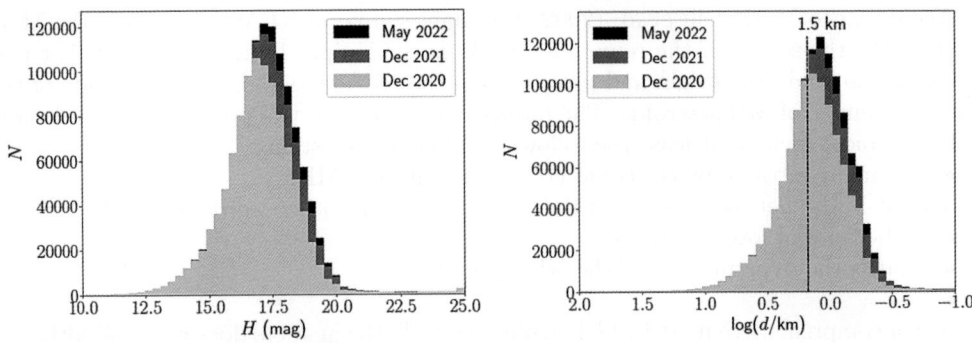

Figure 2. Progress in the knowledge of distribution of MB asteroids by absolute asteroid magnitude H (left) and by size d (right) (data from database of the Minor Planet Center).

NEAs, which is assumed to be statistically complete. It is desirable to have this group as large as possible.

The completeness of the asteriod sample depends on the asteroid size. We believe that for asteroids larger than a certain threshold value the sample can be considered as complete. As it was mentioned in the previous section, to properly study origin and properties of the NEA population which is being permanently replenished from a huge reservoir of MB one needs to model evolution of the entire ensemble of MB asteroids. Again we need to care about completeness of the sample. To estimate the completeness we followed recent changes in size distributions of known MB asteroids. In Fig. 2 the distribution of MB asteroids by absolute asteroid magnitude H and by size is presented. Data were taken from the website of the Minor Planet Center. The dimension (diameter d) of asteroid was determined as in Harris & Harris (1997):

$$d(km) = 10^{3.1236 - 0.5 \log_{10}(A) - 0.2H} \qquad (2)$$

when the albedo value is assumed to be $A = 0.15$.

It is obvious that increase in the number of asteroids discovered is mainly due to smaller asteroids. Almost all larger asteroids have already been discovered. The limit of "completeness of detection" of MB asteroids according to Fig. 2 is achieved at $d = 1.5$ km. However it would be quite problematic to include all MB asteroids larger than ~ 1.5 km into integration of dynamics because of their huge number. We somewhat simplified the task and do not consider the whole ensemble of MB asteroids, but smaller subgroup whose completeness of discovery has been achieved even for smaller asteriod size.

Fig. 3 illustrates the progress in knowledge of size distribution of the asteroids which perihelion distances $q < 1.72$ AU. The number of such asteroids is 10 000. As can be seen from Fig. 3, the size limit of completeness of detection for such asteroids is approximately $d = 800$ m.

Taking into account the requirements of the completeness of the sample and limited computing resources, we studied dynamical evolution of 10 000 asteroids larger that 1 km with $q < 1.72$ AU on a time interval of 20 Myr. Note that 833 NEAs larger than 1 km are included and this NEA sample seems to be complete.

We refer to results for NEA subset as Model 1 and for whole sample of 10 000 asteroids as Model 2. The initial distribution of the studied asteroids in the plane of orbital elements $a - e$ is shown in Fig. 4

Integration of asteroids motion was carried out with the REBOUND software package using the hybrid scheme MERCURIUS Rein et al. (2019). We take into consideration the gravitational field of planets and the Sun. In calculations of the sink rates of asteroids

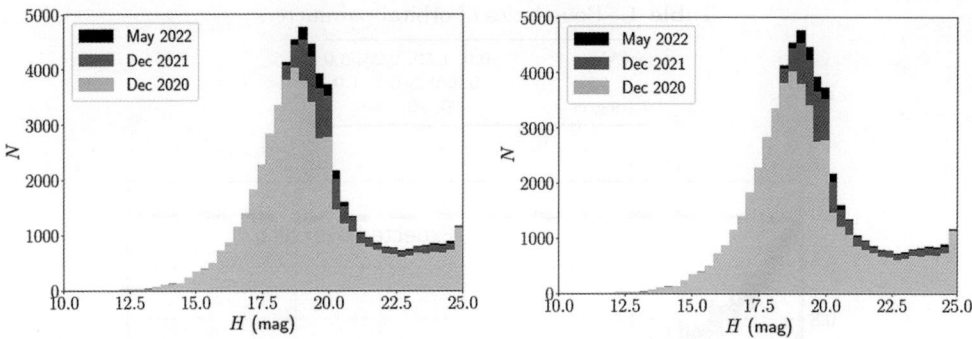

Figure 3. Progress in the knowledge of distribution of MB asteroids with perihelion distances $q < 1.72$ AU.

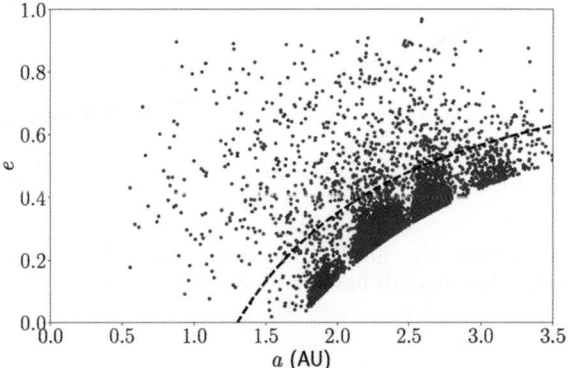

Figure 4. The initial distribution of the studied asteroids larger than 1 km in the plane of the orbital elements $a - e$.

colliding the Sun and planets, we took the geometric dimensions of the objects and neglected sizes of asteroids.

To specify dependence $t_{NEA}(a, e, i)$ the entire area of the orbital parameters of the NEA population was divided into cells. Boundary values for cells are given in Tabl.1.

The number of cells is 18. This is a rather rough grid but because of the relatively low total number of NEAs under consideration it does not seem reasonable to increase the number of cells since number of NEAs in a cell would become too small for statistical estimates. The evolution of the ratio N/N_0, i.e. the ratio of the current number of NEAs to the initial number of NEAs in this cell was traced for each cell, and, accordingly, the value of t_{NEA} was determined for each cell.

To illustrate the role of asteroid-planet approaches, the number of approaches of each asteroid to planets was calculated for all asteroids. The approach is considered as an entry event of an asteroid into a zone wich is three times larger than the Hill sphere of the planet. Also in the Model 1 the statistics of NEA loss through various channels (collisions with the Sun, collisions with planets, asteroids leaving the Solar System, asteroids leaving the NEA scope of definition) was followed.

3. Results

In Fig. 5 a time dependence of the ratio of the number of asteroids remained in the NEA zone to the initial number of NEAs at $t = 0$, i.e. $N(t)/N_0$ for Model 1 and Model 2

Table 1. Boundaries of orbital parameters

a (AU)	0.0, 1.50, 2.25, 3.0
e	0.0, 0.2, 0.7, 1.0
i (degrees)	0, 20, 180

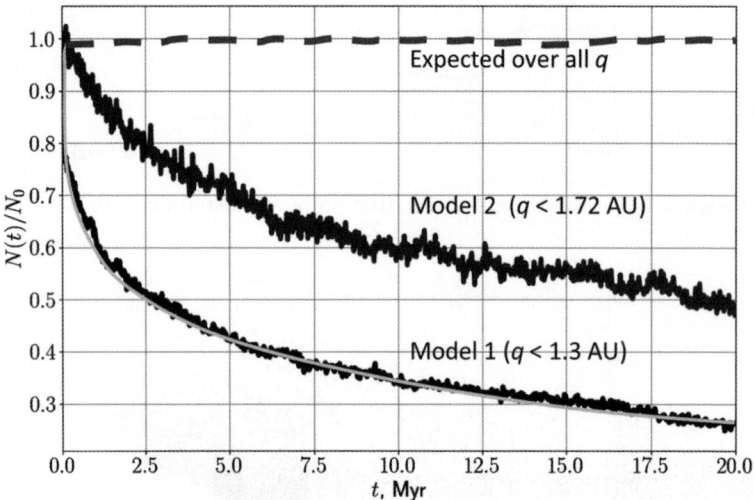

Figure 5. The ratio $N(t)/N_0$ for the NEA ensemble (Model 1) and for the extended set of asteroids (Model 2). The smooth line represents the approximation (3) for Model 1.

are shown. For Model 1 $t_{NEA} \approx 3.0$ million years. The dependence $N(t)/N_0$ in Model 1 is well approximated by the expression

$$\frac{N(t)}{N_0} = e^{-0.5t^{0.33}}, \tag{3}$$

where the time t is given in million years.

Fig. 5 also shows the dependence $N(t)/N_0$ in the Model 2. This value decreases much more slowly over time, which is naturally explained by the fact that there is a significant reservoir of asteroids in the $q < 1.72$ AU zone from which asteroids can "diffuse" into the NEA zone. If one will take into consideration the total reservoir of MB asteroids, then the curve will be flat at the level of $\simeq 1$ (in Fig. 5 is shown by a dashed line).

The dependence of $t_{NEA}(a, e, i)$ is illustrated in Fig. 6.

From Fig. 6 the following conclusions can be drawn:

• the values of t_{NEA} are significantly higher for NEAs whose orbits originally had higher i. Qualitatively this can be explained by the fact that asteroids in these orbits have fewer close approaches to planets.

• the higher values of a and e the lower t_{NEA} is lower for higher values of a and e. Bodies at such orbits have more approaches to the most powerful disturbing body – Jupiter.

It seems to be natural to compare $t_{NEA}(a, e, i)$ and the frequency of approaches N_{enc}. In Fig. 7 numbers of approaches for asteroids from different cells on the orbital parameters plane $a - e$, are given at the time of 12 million years. We checked that the picture does not change even for a larger time interval of integration.

Since the number of approaches can be quite large it is reasonable to split the plane $a - e$ into a larger number of cells in compare to Fig. 6. The upper panel of the figure 7

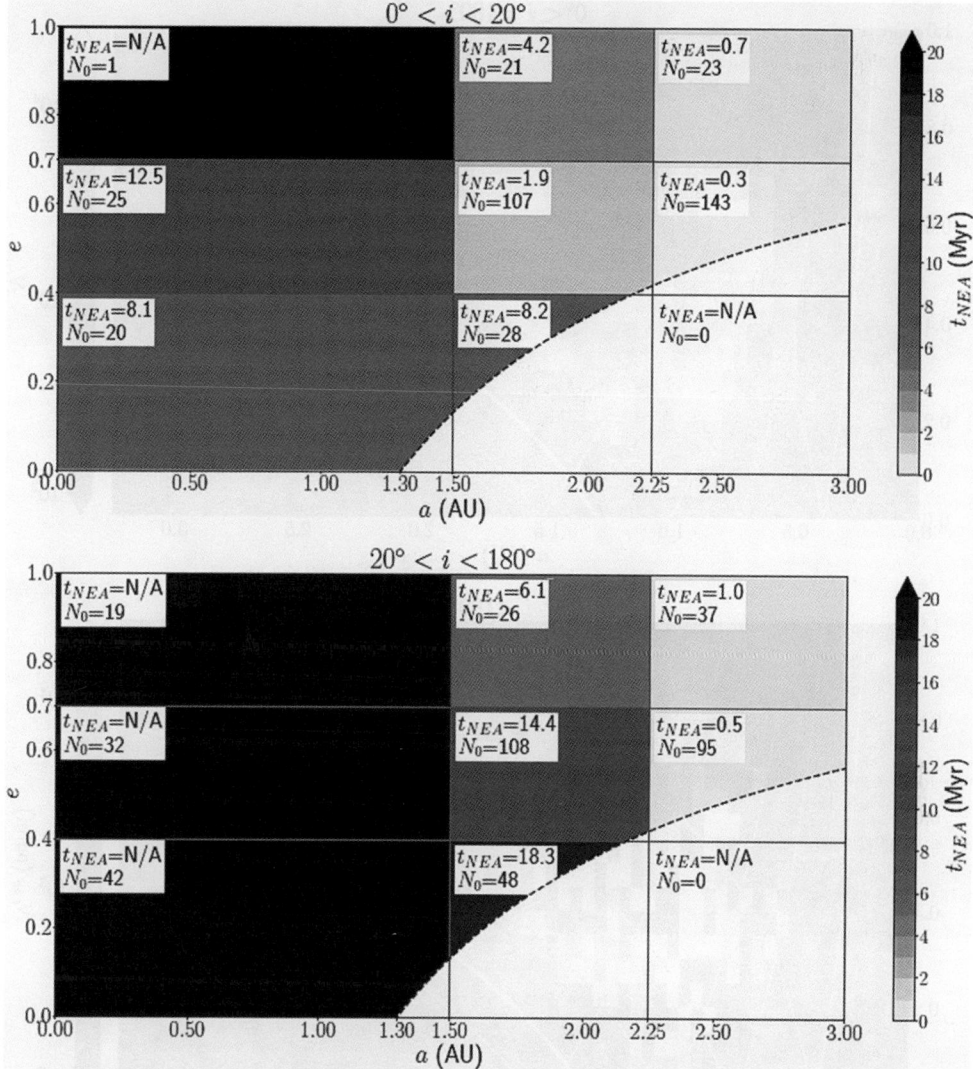

Figure 6. Dependence of $t_{NEA}(a, e, i)$ on the orbital parameters for cells of the parameters. t_{NEA} (in million years) and the number of asteroids N_0 at the time $t_0 = 0$ are indicated for each cell. In black-colored zones, t_{NEA} exceeds 20 million years.

shows the total number of approaches to the planets of asteroids located in this cell. At the lower panel, this value is normalized to average number of asteroids in this cell, i.e. the panel shows an average number of approaches for each asteroid from this cell. A prominent detail in Fig. 7 is an increased number of approaches (both general and relative) for asteroids which perihelion distance is $q = 0.72$ AU and $q = 1.0$ AU as well as those with the aphelion distance $Q = 1.0$ AU. This is due to the fact that such asteroids relatively more frequently approach Venus and the Earth respectively. The increased frequency of approaches seen in the upper right corner of the Fig. 7 is explained by the fact that asteroids in such elongated orbits with large a relatively frequently approach Jupiter in the aphelion region of their orbits.

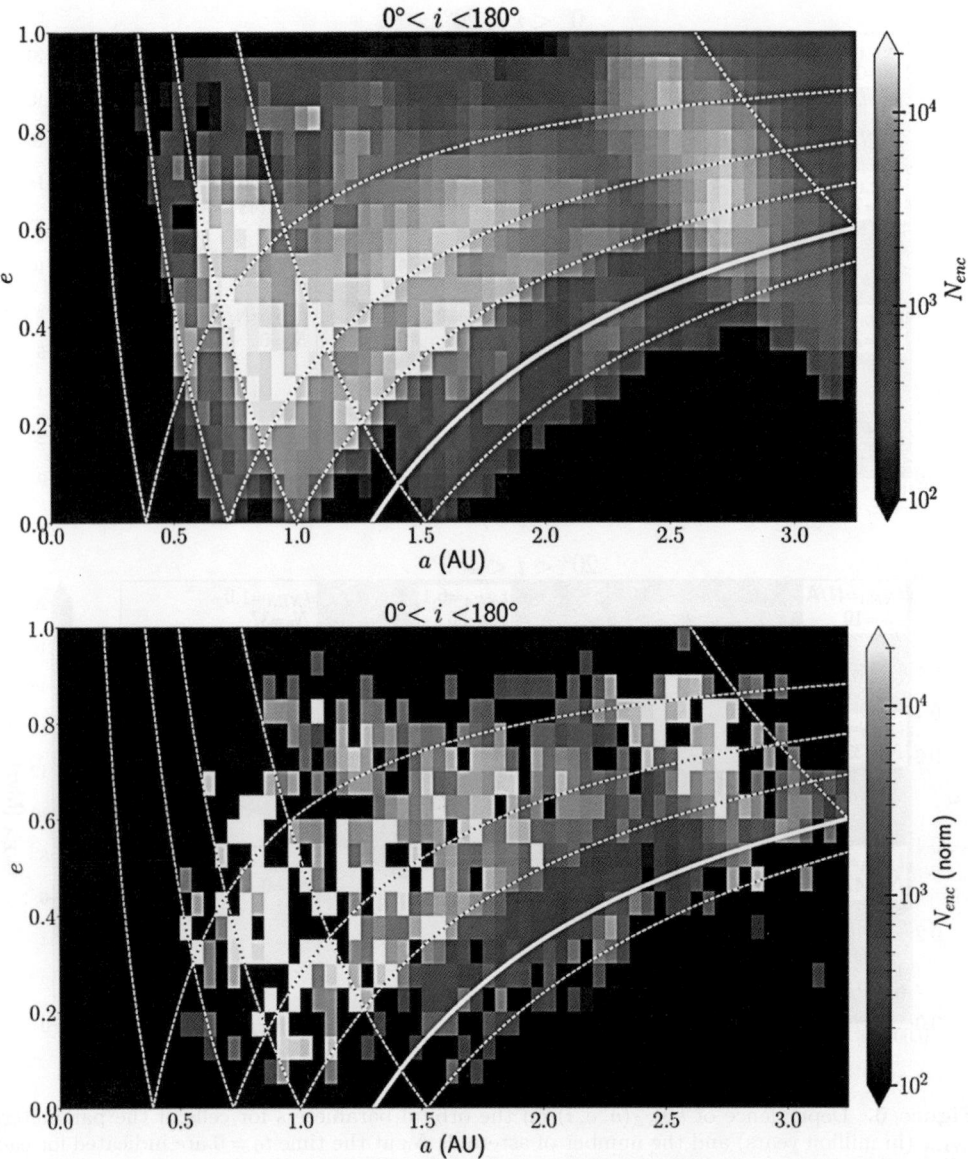

Figure 7. The number of approaches to planets for asteroids from different cells on the plane of the orbital parameters $a - e$ at the time $t = 12$ million years. The top panel shows the total number of approaches for each cell. On the bottom panel - the normalized number of approaches (approaches per asteroid). The dotted lines correspond to the perihelion distance $q = 0.387, 0.723, 1.0, 1.523$ AU, as well as the aphelion distance $Q = 0.387, 0.723, 1.0, 1.523$, 5.203 AU. The bold white line shows the border of the NEO zone ($q = 1.3$ AU).

As it was mentioned above the rates of escape of asteroids from NEA zone via various escape channels were followed. Fig. 8 shows the time dependence of the number $N(t)$ of asteroids (relative to the initial number of N_0) leaving the NEA zone (Model 1). Number of irretrievably lost asteroids (due to ejection from the Solar System, collisions with planets or the Sun) monotonously increases. At $t = 20$ million years the share of NEAs abandoned the Solar System is 7%, collided planets – 4%, collided the Sun – 15%, ejected

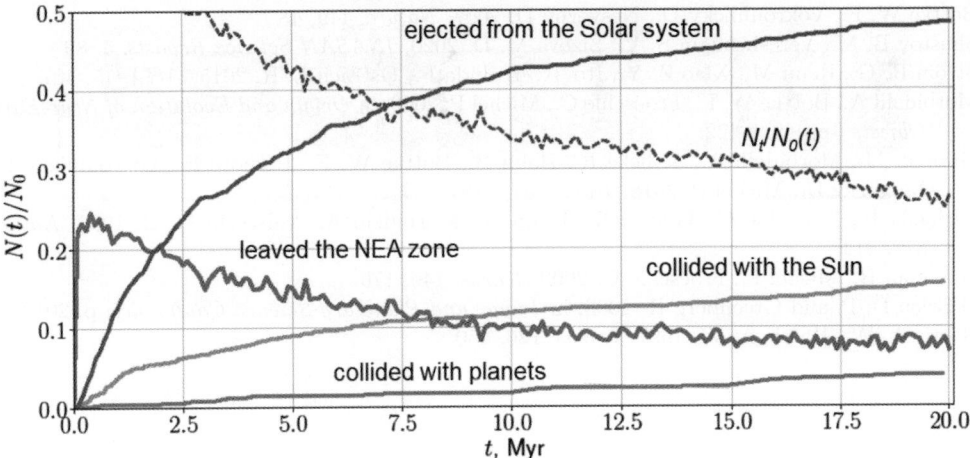

Figure 8. Loss of asteroids from NEA zone via various channels in the process of dynamical evolution (Model 1). $N(t)/N_0$ – ratio of cumulative number of asteroids lost via a given channel at time t to is number of asteroid at $t = 0$. The dashed line is the same dependency as shown in Fig. 4 (Model 1).

from the Solar System – 48%. Major share of collisions occur with the Sun, which was also noted in Farinella et al. (1994).

4. Conclusions

In this paper we quantitatively study the dependence of the loss rate of larger NEAs on the position of asteroids in the space of initial orbital elements (a, e, i). Estimate for the loss time scale (t_{NEA}) is obtained. The main channels of NEAs loss are considered. The results obtained can be used in the future study of mechanisms of replenishment of the NEA population a well as for the study of MB fate.

Dependence of t_{NEA} on initial orbital parameters of NEAs is strong and the t_{NEA} for different regions of the phase space may differ more that order of magnitude. On the other hand the distribution of NEAs in a phase space seems to remain stable for a long time. Hence to consider mechanisms of replenishment one needs to take into account this result as a boundary condition of stationarity. The dependence of t_{NEA} on the asteroid size has not been studied in this work. As shown in Granvik et al. (2018), the dynamic fate of an asteroid of size in range $\simeq 2 - 0.05$ km depends on the size (to a relatively lesser extent then for orbital parameters). Most important factor was considered to be the Yarkovsky effect. In this paper we consider dynamics of larger (> 1 km) asteroids. For these asteroids Yarkovsky effect is relatively weak ($da/dt < 10^{-9}$ AU/yr). In future models we plan to expand the list of NEAs by adding smaller asteroids, but requirement of completeness of the sampe remains. Then this dependence on size is to be taken into account.

The work was supported by the RSCF grant No. 22-12-00115

References

Zolotarev R. V. and Shustov B. M. 2021, *Astron. Rep.*, 65, 518
Rein H., Hernandez D. M., Tamayo D., Brown G., et al. 2019, *MNRAS*, 485, 5490
Neukum G. and Ivanov B. A. 2002, *Lunar and Planetary Science Conference*, 1263
Mazrouei S., Ghent R. R., Bottke W. F., Parker A. H., Gernon T. M. 2019, *Science* 363, 253
Ipatov S. I., Feoktistova E. A., Svettsov V. V. 2020, *Solar System Research*, 54, 384

Bottke W. F., Vokrouhlicky D., Nesvorny D. 2007, *Nature*, 449, 48

Shustov B. M., Vereshchagin S. V., Sizova M. D. 2020, *INASAN Science Reports*, 5, 89

Strom R. G., Renu M., Xiao Z.-Y., Ito T., Yoshida F., Ostrach L. R. 2015, *A&A*, 15, 407

Morbidelli A., Bottke W. F., Froeschle C., Michel P. 2002, in *Origin and Evolution of Near-Earth Objects* , pp. 409–422.

Granvik M., Morbidelli A., Jedicke R., Bolin B., Bottke W. F., Beshore E., Vokrouhlicky D., Nesvorny D., Michel P. 2018, *Icarus*, 312, 181

Farinella P., Froeschle C., Gonczi R., Hahn G., Morbidelli A., Valsecchi G. B. 1994, *Nature*, 371, 314

Gladman B., Michel P., Froeschle C. 2000, *Icarus*, 146, 176

O'Brien D. P. and Greenberg R. 2003, in *Lunar and Planetary Science Conference*, p. 2018

Harris A. W., Harris A. W. 1997, *Icarus*, 126, 450

Astronomical Hazards for Life on Earth
Proceedings IAU Symposium No. 374, 2025
G. Tancredi, ed.
doi:10.1017/S1743921324000735

Migration of bodies to the Earth from different distances from the Sun

Sergei I. Ipatov ⓘ

Vernadsky Institute of Geochemistry and Analytical Chemistry of RAS,
119991, 19 Kosygin st., Moscow, Russia
email: siipatov@hotmail.com

Abstract. Migration of bodies under the gravitational influence of almost formed planets was studied, and probabilities of their collisions with the Earth and other terrestrial planets were calculated. Based on the probabilities, several conclusions on the accumulation of the terrestrial planets have been made. The outer layers of the Earth and Venus could accumulate similar planetesimals from different regions of the feeding zone of the terrestrial planets. The probabilities of collisions of bodies during their dynamical lifetimes with the Earth could be up to 0.001-0.01 for some initial semi-major axes between 3.2 and 3.6 AU, whereas such probabilities did not exceed 10^{-5} at initial semi-major axes between 12 and 40 AU. The total mass of water delivered to the Earth from beyond Jupiter's orbit could exceed the mass of the Earth's oceans. The zone of the outer asteroid belt could be one of the sources of the late-heavy bombardment. The bodies that came from the zone of Jupiter and Saturn typically collided with the Earth and the Moon with velocities from 23 to 26 km/s and from 20 to 23 km/s, respectively.

Keywords. methods: n-body simulations, Earth, Moon, solar system: formation

---

1. Introduction

Migration processes played a great role in accumulation of planets and in delivery of water to the Earth and the terrestrial planets. The life on the Earth could not appear without water. However, collisions of bodies with the Earth can be dangerous for life and can kill some organisms, including people. The Earth's ocean water and its D/H ratio could be the result of mixing water from several exogenous and endogenous sources with high and low D/H ratio. It was considered that bodies containing ices migrated to the Earth from the outer part of the Main asteroid belt (O'Brien et al. 2014; Morbidelli et al. 2000; Morbidelli et al. 2012; Petit et al. 2001; Raymond et al. 2004; Lunine et al. 2003) and from beyond Jupiter's orbit (Morbidelli et al. 2000; Levison et al. 2001; Ipatov & Mather 2004; Ipatov & Mather 2006; Ipatov 2010; Marov & Ipatov 2018).

The studies of the process of accumulation of the terrestrial planets were often (e.g,. Chambers & Wetherill 1998; Ipatov 1993; Chambers 2013; Raymond et al. 2009) based on computer simulations of the evolution of disks of gravitating bodies combined at their collisions. In (Ipatov 1993) initial planetesimals were divided into four groups depending on their semi-major axes. It was obtained that composition of the largest formed planets was similar and was close to the composition of the initial disk.

2. Initial data and considered models

Below migration of bodies-planetesimasl is considered at the stage when planets already have got their masses and orbits. Such migration corresponds to the late stages of accumulation of planets and to the migration of bodies in the present Solar System. The main attention is paid to the probabilities of collisions of bodies with the Earth and the Moon. The studies were based on the computer N-body simulations of migration of bodies-planetesimals under the gravitational influence of planets. For integration of the motion equations, the symplectic method of the Swift integration package (Levison & Duncan 1994) was used. The bodies-planetesimals that collided with the Sun or reached 2000 AU were excluded from integration.

Two models were considered. In model C, bodies-planetesimals that collided with planets were excluded from integration. This model was used for studies of the number of collisions of bodies-planetesimals initially located in the zone of the terrestrial planets. For studies of migration to the Earth of the bodies-planetesimals initially located in the zone of the giant planets and in the outer asteroid belt, in model MP at integration, planets were considered as material points, and the probabilities of collisions of bodies with planets were calculated based on the arrays of the orbital elements of migrated bodies during the considered time interval similar to (Ipatov & Mather 2004; Ipatov & Mather 2006; Ipatov 2019). Such probabilities are relatively small. This model MP allows to calculate probabilities of collisions of bodies-planetesimals with planets in the case of small probabilities without considering a large number of bodies. The probability was averaged over all considered bodies-planetesimals, and the probability per one body was considered. Note that this algorithm of calculations of the probabilities differed from the Öpik's scheme (used, for example, by Morbidelli et al. (2000) and Nesvorny et al. (2017)). It included calculations of a synodic period and the region where the distance between the 'first' orbit and the projection of the 'second' orbit onto the plane of the 'first' orbit is less than the sphere of action (i.e., the Tisserand sphere).

In each variant of calculations, 250 initial bodies-planetesimals were considered. The semi-major axes a_o of the initial orbits of the bodies varied from $a_{\min}$ to $a_{\min} + d_a$. The number of bodies with a_o was proportional to $a_o^{0.5}$. For $(i+1)$-th body-planetesimal, the value of a_o was calculated with the use of the formula $a_o(i+1) = (a_{oi}^2 + [(a_{\min} + d_a)^2 - a_{\min}^2]/N_o)^{1/2}$, where a_{oi} is the value of a_o for i-th body-planetesimal, and $N_o = 250$. Initial eccentricities and inclinations of orbits of bodies equaled to e_o and $i_o = e_o/2$ rad, respectively.

Three main series of calculations were considered. In series Ter, bodies-planetesimals in the zone of the terrestrial planets were considered. In this case $d_a = 0.2$ AU, exclusive for $a_{\min} = 1.5$ AU (when $d_a = 0.5$ AU). For different variants of Ter calculations, the values of $a_{\min}$ varied with a step of 0.2 AU from 0.3 to 1.5 AU. e_o was equaled to 0.05 or 0.3. In series AsB, initial bodies were located in the outer asteroid belt, $d_a = 0.1$ AU, $a_{\min}$ varied from 3 to 4.9 AU, and e_o equaled to 0.02 or 0.15. In series JN, initial bodies were located in the zone of the giant planets, $d_a = 2.5$ AU, $a_{\min}$ varied from 2.5 to 40 AU, and e_o equaled to 0.05 or 0.3. In series Ter, the gravitational influence of the Sun and all planets was taken into account. In series AsB and JN, the gravitational influence of Mercury was not considered.

3. Mixing of planetesimals in the feeding zone of the terrestrial planets

In (Ipatov 2019) the probability of a collision of a planetesimal with a planet was calculated for the MP model based on the arrays of orbital elements of bodies-planetesimals during the considered time interval. For a few planetesimals from the feeding zone of the terrestrial planets such calculated probabilities exceeded 1. Based on the arrays of orbital elements used in (Ipatov 2019), I recalculated the probabilities of collisions of planetesimals with planets for the model in which calculations of the probability p_i of a collision of each i-th planetesimal with a planet stopped when the probability reached 1. Though the recalculated probabilities could be smaller than those in (Ipatov 2019), the conclusions about the mixing of planetesimals were similar. The probabilities of collisions of planetesimals with planets could be even smaller for C calculations that exclude collided planetesimals from integration.

Below I present the probabilities of collisions of planetesimals with planets obtained at the C calculations that exclude collided planetesimals from integration of their motion and consider present planets. For $e_o = 0.05$ during the dynamical lifetimes T_{end} of planetesimals, the fraction p_E of initial planetesimals collided with the Earth was about 0.35 at a_{min} equal 0.9 AU, 0.2 at a_{min} equal 0.7 or 1.1 AU, and 0.1-0.15 at a_{min} equal to 0.5, 1.3, or 1.5 AU. For $e_o = 0.3$, the values of p_E were about 0.15-0.2 for $0.7 \leqslant a_{min} \leqslant 1.3$ AU, and about 0.1 at a_{min} equal to 0.5 or 1.5 AU. The ratio of the number of planetesimals collided with the Earth during the first $T = 1$ Myr to that collided during T_{end} exceed 0.5 at $a_{min} = 0.9$ AU and $e_o = 0.05$. The similar ratio at $T = 10$ Myr was between 0.4 and 0.7 at a_{min} equal to 0.5, 1.1 and 1.3 AU for e_o equal to 0.05 or 0.3. The values of T_{end} could exceed 1000 or 2000 Myr for $e_o = 0.05$ and $0.5 \leqslant a_{min} \leqslant 1.1$ AU. For other considered variants, they typically equaled to several hundreds of million years. The fraction p_V of planetesimals collided with Venus exceed p_E usually at all considered variants exclusive for $a_{min} = 0.9$ AU and $e_o = 0.05$ and for $a_{min} = 1.5$ AU. For the latter variants, the ratio p_V/p_E was about 0.7-0.9 at T_{end}. The ratio p_V/p_E was about 4-6, 2-3, 1.3-1.4, and 1.6-1.8 for a_{min} equal to 0.5, 0.7, 1.1, and 1.3 AU, respectively (both at e_o equal to 0.05 or 0.3). The above results testify that more than a half of planetesimals with initial semi-major axes between 0.9 and 1.1 AU and initial eccentricities $e_o = 0.05$ that collided with the Earth of its present mass collided in less than 1 Myr. The Earth embryo first mainly accumulated planetesimals that moved not far from its orbits. Planetesimals that initially had been located near orbits of other planets could collide with the Earth after tens or even hundreds of million years after the beginning of the evolution. At the late stages of accumulation of Earth and Venus, both planets accumulated planetesimals originated at the same distance from the Sun.

4. Migration of bodies from the zone of the giant planets to the terrestrial planets

Sources of the water delivered to the Earth included bodies-planetesimals migrated from the region of the outer asteroid belt and from the region beyond the Jupiter's orbit. Migration of bodies for different values of a_{min} from 2.5 to 40 AU was studied for JN series (Ipatov et al. 2020). In the Solar System at $a_{min} \leqslant 10$ AU, the values of the probability p_E of a collision of a planetesimal with the Earth averaged over 250 bodies can vary by up to a factor of a thousand for different calculation variants with 250 planetesimals, and in some calculations they exceeded 10^{-3}. This large difference can be caused by that in some calculations there was a planetesimal, for which the probability of a collision with the Earth was greater than for all other 249 planetesimals because one of thousands of planetesimals could get the Earth-crossing orbit and move in it for millions years. For comparison, the mean time for motion of a former Jupiter-crossing object in Earth-crossing orbit is about 30 Kyr.

While averaging over thousands of planetesimals, at $5 \leqslant a_o \leqslant 12$ AU, p_E could exceed $2 \cdot 10^{-6}$ by at least a factor of several. On average, for a planetesimal at $20 \leqslant a_o \leqslant 40$ AU, p_E was about 10^{-6}. At $p_E = 2 \cdot 10^{-6}$ and the total mass of planetesimals of about $100 m_E$ (where m_E is the mass of the Earth), the total mass of planetesimals collided with the Earth equaled to about the mass of the Earth's oceans ($\sim 2 \cdot 10^{-4} m_E$). Some fraction of material delivered from the zone of the giant planets to the Earth was composed of water and volatiles. In comets such fraction does not exceed 1/3 (e.g., Davidsson et al. 2016; Fulle et al. 2017). However, some authors suppose that primary planetesimals could contain more ice than it is now found in comets. Lodders (2003), Howard et al. (2014), and Ciesla et al. (2015) supposed that solids beyond the water line should be 50 per cent water in mass. The total mass of water delivered to the Earth from beyond Jupiter's orbit could exceed the mass of the Earth's oceans if the total mass of planetesimals was about $200 m_E$. The ratio of the total mass of the material delivered from beyond the orbit of Jupiter to a planet to the mass of the planet was about two times greater for Mars than that for the Earth, and such ratios for Mercury and Venus were a little greater than that for the Earth. Some bodies from the zone of Uranus and Neptune may have fallen onto the Earth within more than twenty million years.

5. Migration of bodies from the zone of the outer asteroid belt to the terrestrial planets

For the series AsB (with $d_a = 0.1$ AU), a_{min} varied from 3 to 4.9 AU; $e_o = 0.02$ and $e_o = 0.15$ were taken. At time interval $T = 100$ Myr, the values of the probability p_E of a collision of a planetesimal with the Earth averaged over 250 planetesimals vary from less than 10^{-6} to values of the order of 10^{-3} (and of 0.01 at $T = 1000$ Myr) at $a_{min} < 4$ AU (Ipatov 2021). Generally, the values of p_E are often between 10^{-6} and 10^{-5} at $a_{min} \geqslant 4.1$ AU, as for calculations with $5 \leqslant a_{min} \leqslant 10$ AU. Calculations with 250 planetesimals with $p_E > 2 \cdot 10^{-5}$ were obtained more often at $e_o = 0.02$ for $3.2 \leqslant a_{min} \leqslant 3.3$ AU and $a_{min} = 3.5$ AU (at $e_o = 0.15$ also for $3.8 \leqslant a_{min} \leqslant 3.9$ AU). In some variants of calculations at $a_{min} = 3.3$ AU and $a_{min} = 3.5$ AU for $e_o = 0.02$, and at a_{min} equal to 3.0, 3.2, 3.3 and 3.8 AU for $e_o = 0.15$, the value of p_E was greater than 10^{-3} for a time interval up to a few billion years. Most of collisions with the Earth of bodies at $4 \leqslant a_o \leqslant 5$ AU occurred during the first 10 Myrs. Bodies-planetesimals that initially crossed the Jupiter's orbit could reach the Earth's orbit mostly within the first Myr after the formation of Jupiter. For $3 \leqslant a_{min} \leqslant 3.5$ AU and $e_o \leqslant 0.15$, some bodies could fall onto the Earth and the Moon in a few billions years. For example, $p_E = 4 \cdot 10^{-5}$ for $a_{min} = 3.3$ AU, $e_o = 0.02$ at $500 \leqslant t \leqslant 800$ Myr, and $p_E = 6 \cdot 10^{-6}$ at $2000 \leqslant t \leqslant 2500$ Myr. The zone of the outer asteroid belt can be one of the sources of the late heavy bombardment.

6. Probabilities of collisions of migrated bodies with the Moon

During accumulation of the terrestrial planets, the ratio of the number of planetesimals from the feeding zone of these planets colliding with the Earth to that colliding with the Moon varied mainly from 20 to 40. For bodies arriving from distances from the Sun greater than 3 AU, this ratio was close to 17. The ratio of the total mass of planetesimals collided with a celestial object to the mass of the object was greater for the Moon than for the Earth. However, the fraction of the material of collided planetesimals that was left in a celestial object was greater for the Earth than that for the Moon.

The characteristic velocities of collisions of planetesimals from the feeding zone of the terrestrial planets with the Moon varied mostly from 8 to 16 km/s, and velocities of collisions with the Earth were from 13 to 19 km/s, depending on the initial distances of planetesimals from the Sun and on their initial eccentricities (Marov & Ipatov 2021). The velocities of collisions with the Moon of bodies that came from the feeding zones of Jupiter and Saturn were mainly from 20 to 23 km/s. For the Earth this diapason was from 23 to 26 km/s. The characteristic velocities of collisions of the planetesimals, originally located at a distance from the Sun from 0.7 to 1.1 AU, with the embryos of the Earth and the Moon with masses 10 times less than the present masses of these celestial objects, were mainly in the range from 7 to 8 km/s for the embryo of the Earth and from 5 to 6 km/s for the embryo of the Moon.

7. Time variations in the number of 1-km near-Earth objects

Based on analysis of lunar craters, Mazrouei et al. (2019) concluded that the probability of collisions of near-Earth objects (NEOs) with the Moon increased 2.6 times 290 Myr ago. For the model, in which the probability of a collision of a NEO with the Moon was equal to the current value for the last 290 Myr, and before that within 810 Myr it was 2.6 times less than the current value, the number of formed craters would be 0.6 from (i.e., it would be 1.7 times less than) the estimate obtained on the basis of the current number of NEOs. Ipatov et al. (2020) compared the number of lunar craters larger than 15 km across and younger than 1.1 Ga with the estimates of the number of craters that could have been formed for 1.1 Ga if the number of near-Earth objects and their orbital elements during that time were close to the corresponding current values. These estimates do not contradict to the growth in the number of near-Earth objects after probable catastrophic fragmentations of large main-belt asteroids, which may have occurred over the recent 300 Myr; however, they do not prove this increase. For some models, the cratering rate may have been constant over the recent 1.1 Ga. The estimates made in (Ipatov et al. 2020) allow an increase in the probability of collisions of NEOs with the Moon by a factor of 2.6 about 290 Myr ago. With this conclusion, the paper by Mazrouei et al. (2019) agrees better with the estimates based on the number of craters per unit area for the region of the Ocean of Storm and other seas of the visible side of the Moon. It was assumed that the number of Copernican craters per unit area for the entire surface of the Moon could be approximately the same as that for the region of the seas, that is, be more than the current estimate for the entire surface of the Moon.

8. Conclusions

The outer layers of the Earth and Venus could accumulate similar planetesimals from different regions of the feeding zone of the terrestrial planets. The total mass of water delivered to the Earth from beyond Jupiter's orbit could exceed the mass of the Earth's oceans. The zone of the outer asteroid belt could be one of the sources of the late heavy bombardment. The bodies that came from the zone of Jupiter and Saturn typically collided with the Earth and the Moon with velocities from 23 to 26 km/s and from 20 to 23 km/s, respectively.

9. Acknowledgements

Studies of the migration of planetesimals to the terrestrial planets were carried out as a part of the state assignments of the Vernadsky Institute of RAS. The studies of the migration of planetesimals to the Moon were supported by the grant of Russian Science Foundation N 21-17-00120.

References

Chambers, J. 2013, *Icarus*, 224, 43

Chambers, J. & Wetherill, G.W. 1998, *Icarus*, 136, 304

Ciesla, F.J., Mulders, G.D., Pascucci, I., & Apai, D. 2015, *ApJ* 804, 9 (11 pp.). DOI:10.1088/0004-637X/804/1/9

Davidsson, B.J.R., Sierks, H., Guttler, C., et al. 2016, *A& A* 592, A63

Fulle, M., Della Corte, V., Rotundi, A., et al. 2017, *MNRAS*, 469, S45

Howard, K.T., Alexander, C.M.O'D., & Dyl, K.A. 2014, *Lunar Planet. Sci.* 45, Abstract 1830

Ipatov, S.I. 1993, *Solar System Research*, 27, 83, https://www.academia.edu/44448077

Ipatov, S.I. 2010, in: Fernandez, J.A., Lazzaro, D., Prialnik, D., & Schulz, R (eds.), *Proc. Int. Astron. Union*, Symp. S263. "Icy Bodies in the Solar System", (Cambridge Univ. Press), p. 41. https://arxiv.org/abs/0910.3017

Ipatov, S.I. 2019, *Solar System Research*, 53, 332, DOI: 10.1134/S0038094619050046, https://arxiv.org/abs/2003.11301

Ipatov, S.I. 2020, *EPSC abstracts*, EPSC2020-71, DOI: 10.5194/epsc2020-71

Ipatov, S.I. 2021, *EPSC abstracts*, EPSC2021-100, DOI: 10.5194/epsc2021-100

Ipatov, S.I. & Mather, J.C. 2004, *Annals of the New York Academy of Sciences* 1017, 46, DOI: 10.1196/annals.1311.004, https://arxiv.org/format/astro-ph/0308448

Ipatov, S.I. & Mather, J.C. 2006, *Adv. Sp. Res.* 37, 126, DOI: 10.1016/j.asr.2005.05.076, https://arxiv.org/abs/astro-ph/0411004

Ipatov, S.I., Feoktistova, E.A., & Svetsov, V.V. 2020, *Solar System Research*, 54, 384. DOI: 10.1134/S0038094620050019, https://arxiv.org/abs/2011.00361

Levison, H.F. & Duncan, M.J. 1994, *Icarus* 108, 18

Levison, H.F., Dones, L., Chapman, C.R., et al. 2001, *Icarus* 151, 286

Lodders, K. 2003, *ApJ* 591, 1220

Lunine, J.I., Chambers, J., Morbidelli, A., & Leshin, L.A. 2003, *Icarus*, 165, 1

Marov, M.Ya. & Ipatov, S.I. 2018, *Solar System Research*, 52, 392, DOI: 10.1134/S0038094618050052, https://arxiv.org/abs/2003.09982

Marov, M.Ya. & Ipatov, S.I. 2021, *Geochemistry International*, 59, 1010. DOI: 10.1134/S0016702921110070, https://arxiv.org/abs/2112.06047

Mazrouei, S., Ghent, R.R., Bottke, W.F., et al. 2019, *Science*, 363, 253

Mezger, K., Debaille, V., & Kleine, T. 2013, *Space Sci. Revs.* 174, 27

Morbidelli, A., Chambers, J., Lunine, J.I., et al. 2000, *Meteor. Planet. Sci.* 35, 1309

Morbidelli, A., Lunine, J.I., O'Brien, D.P., et al. 2012, *Ann. Rev. Earth Planet. Sci.* 40, 251

Nesvorny, D., Roig, F., & Bottke, W.F. 2017, *AJ*, 153, A103

O'Brien, D.P., Walsh, K.J., Morbidelli, A., et al. 2014, *Icarus*, 239, 74

Ormel, C.W., Liu, B., & Shoonenberg, D. 2017, *A& A.*, 604, A1

Petit, J.-M., Morbidelli, A., & Chambers, J. 2001, *Icarus*, 153, 338

Raymond, S.N., Quinn, T., & Lunine, J.I. 2004, *Icarus*, 168, 1

Raymond, S.N., O'Brien, D.P., Morbidelli, A., & Kaib, N.A. 2009, *Icarus*, 203, 644

Astronomical Hazards for Life on Earth
Proceedings IAU Symposium No. 374, 2025
G. Tancredi, ed.
doi:10.1017/S1743921324000760

Dynamic and physical parameters of near-Earth asteroids from the observations

Eduard Kuznetsov[1], **Dmitry Glamazda[1]**, **Galina Kaiser[1]**,
Vadim Krushinsky[1], **Sergey Kryuchkov[2]**, **Sergey Naroenkov[2]**,
Alexander Perminov[1] and **Yulia Wiebe[1]**

[1]Kourovka Astronomical Observatory, Ural Federal University,
Lenina Avenue, 51, Yekaterinburg, 620000, Russia
email: `Eduard.Kuznetsov@urfu.ru`

[2]Institute of Astronomy of the Russian Academy of Sciences,
Pyatnitskaya Street, 48, Moscow, 119017, Russia
email: `snaroenkov@gmail.com`

Abstract. We performed astrometric and multicolor photometric observations of near-Earth asteroids at the SBG telescope of the Kourovka Astronomical Observatory of the Ural Federal University and the Zeiss-1000 telescope of the Simeiz Observatory of the Institute of Astronomy of the Russian Academy of Sciences. We improved orbital elements and estimated the A_2 acceleration due to the Yarkovsky effect for asteroids (52768) 1998 OR2, (65690) 1991 DG, (159857) 2004 LJ1, (326732) 2003 HB6, (332446) 2008 AF4, (388945) 2008 TZ3, 2015 NU13 from astrometric observations with the SBG telescope. Furthermore, we estimated the axial rotation periods of the asteroids (137170) 1999 HF1, (159857) 2004 LJ1, (326732) 2003 HB6 from photometric observations with the SBG telescope. We obtained color indices for the asteroids (137170) 1999 HF1, (138127) 2000 EE14, (153591) 2001 SN263, (159857) 2004 LJ1, (326732) 2003 HB6, 2010 TV149 from multicolor photometric observations with the Zeiss-1000 telescope. Furthermore, we estimated the taxonomic classes for three asteroids, according to the color indices: the asteroid (153591) 2001 SN263 has class C, (159857) 2004 LJ1 has class S, and (326732) 2003 HB6 has class D.

Keywords. Near-Earth asteroids, Yarkovsky effect, rotation period, color indices, taxonomic classes

1. Introduction

The Kourovka Astronomical Observatory of the Ural Federal University (AO UrFU) and the Simeiz Observatory of the Institute of Astronomy of the Russian Academy of Sciences (INASAN) carry out a joint project to determine the dynamic and physical parameters of near-Earth asteroids (NEA) based on the results of astrometric and photometric observations. The following tasks are being solved.

- Determination of improved orbit elements.
- Estimation of the non-gravitational acceleration A_2 due to the Yarkovsky effect.
- Determination of the period of axial rotation.
- Estimation of color indices.
- Determining the taxonomic type of asteroids.

We used the SBG telescope of AO UrFU and Zeiss-1000 telescope of the Simeiz Observatory of INASAN (Table 1) to carry out NEA observations. The SBG telescope (Fig. 1) performs astrometric observations in the R filter and photometric observations in

Table 1. Specification of telescopes.

Specification	SBG	Zeiss-1000
Type	Schmidt telescope	Richie-Chrétien telescope
Diameter [m]	0.4	1.0
Focal length [m]	0.8	13
		5.7 (with focus reducer)
CCD camera	Apogee Alta U32	FLI PL 16803
Pixel size [μm]	6.8×6.8	9×9
Image scale [arcsec pixel^{-1}]	1.8	0.143
		0.326 (with focus reducer)
Field of view [arcmin]	65×44	10×10
		22×22 (with focus reducer)

Figure 1. The SBG telescope.

the V and R filters. The Zeiss-1000 telescope (Fig. 2) performs photometric observations in the B, V, R, and I filters.

Table 2 provides information on observations made in 2021–2022. The results obtained from these observations are described in the article. Section 2 presents the results of positional observations. In Section 3, the periods of axial rotation of asteroids are obtained. Section 4 is devoted to the evaluation of color indices. Taxonomic types of asteroids are discussed in section 5.

2. Determination of improved elements of orbits and non-gravitational acceleration A_2

Table 3 contains a list of asteroids observed with the SBG telescope and results of astrometric reductions. Observations of the potentially hazardous asteroid (99942) Apophis were carried out as part of an international campaign (Reddy *et al.* (2022))

Figure 2. The Zeiss-1000 telescope.

organized by the IAWN†. We performed astrometric reduction using the IzmCCD‡ software (Izmailov *et al.* (2010)) for all frames obtained using astrometric and photometric observation programs. As can be seen from Table 3, the root-mean-square errors of the equatorial coordinates of the σ_α and σ_δ asteroids do not exceed 0.3″.

Based on the results of astrometric observations using the IDA software (Galushina *et al.* (2019)), we improved the orbits of asteroids that were observed with the SBG telescope in 2020–2021. For 7 asteroids, estimates of the A_2 acceleration were obtained (Table 4), which characterizes the influence of the Yarkovsky effect. We estimated the A_2 acceleration with an accuracy of 1 to 3 standard deviations, which will allow them to be used in modeling the dynamic evolution of asteroids.

3. Determination of periods of axial rotation

The initial processing of photometric observations was performed using the AM:PM software (Krushinsky (2018)). We used two- or three-parameter models (Muinonen *et al.* (2010); Oszkiewicz *et al.* (2011); Penttilä *et al.* (2016)) to take into account the dependence of the asteroid's brightness on the phase angle. We applied the "Online calculator for H, G_1, G_2 photometric system"¶ to estimate an absolute magnitude H and slope-parameters G_1, G_2 or G. We construct phase light curves in the "Period search service"‖ online system using the Lafler & Kinman (1965) method.

Figure 3 shows the phase light curve of the asteroid (159857) 2004 LJ1, constructed from the results of photometric observations in the V and R filters. The red curve in Fig. 3

---

† International Asteroid Warning Network, https://iawn.net/obscamp/Apophis/
‡ http://izmccd.puldb.ru/
¶ http://h152.it.helsinki.fi/HG1G2/
‖ http://scan.sai.msu.ru/lk/

Table 2. Observation data.

Telescope	Astrometric observations			Photometric observations			Filters
		Number of			Number of		
	NEA	nights	frames	NEA	nights	frames	
SBG	6	17	533	6	39	2625	V, R
Zeiss-1000	—	—	—	6	17	1370	B, V, R, I

Table 3. Data on astrometric processing of observations of asteroids.

	Observations in V and R filters				Astrometric reducing		
NEA	Observation interval [day]	Number of nights	$N_{ph}(R, V)+$ $+N_a(R)$	$N_d(R, V)$	σ_α [arcsec]	σ_δ [arcsec]
(99942) Apophis	62	14	503	387	0.27	0.25
(153591) 2001 SN263	4	2	295	239	0.15	0.20
(159857) 2004 LJ1	36	13	1052	1016	0.22	0.20
(326732) 2003 HB6	6	5	588	583	0.10	0.11
(332446) 2008 AF4	30	5	472	243	0.19	0.15
2015 NU13	1	1	60	60	0.17	0.15

Notes:
$N_{ph}(R, V)$ are the number of frames obtained during photometric observations in the R and V filters;
$N_a(R)$ is number of frames obtained during astrometric observations in the R filter;
$N_d(R, V)$ is number of asteroid positions determined as a result of astrometric reducing;
σ_α and σ_δ are root-mean-square errors of asteroid coordinates in right ascension and declination.

Table 4. A_2 acceleration.

NEA	Epoch	JD_1	JD_2	N	N_{168}	A_2 $[10^{-14}$ au d$^{-2}]$	σ_{A_2} $[10^{-14}$ au d$^{-2}]$
(52768) 1998 OR2	2458944.5	2454834.5	2459107.5	5238	38	−2.71	0.69
(65690) 1991 DG	2458298.5	2452898.5	2459169.5	1096	7	4.22	0.99
(159857) 2004 LJ1	2459405.5	2457035.5	2459487.5	1236	34	−4.84	3.58
(326732) 2003 HB6	2459444.5	2452724.5	2459545.5	3245	23	−2.67	0.69
(332446) 2008 AF4	2458645.5	2454475.5	2459279.5	995	22	−2.13	0.36
(388945) 2008 TZ3	2458241.5	2457461.5	2459143.5	1182	22	3.01	1.47
2015 NU13	2459225.5	2457215.5	2459291.5	316	5	12.7	5.00

Notes:
JD_1 and JD_2 are Julian start and end dates of observations from the Minor Planet Center database (MPC, https://minorplanetcenter.net/db_search) used to improve the orbit;
N is number of observations on the integration interval from the MPC database;
N_{168} is the number of observations made at the Kourovka Astronomical Observatory of UrFU;
σ_{A_2} is root-mean-square error of A_2 acceleration determination.

corresponds to the moving average. Estimated periods P of axial rotation of asteroids (159857) 2004 LJ1 and (326732) 2003 HB6 are given in Table 5. The results are consistent with the data on asteroid rotation periods P_0, contained in the database of photometric light curves of asteroids of the Minor Planet Center**, within the ranges of errors.

4. Determination of color indices

Table 6 gives estimates of the color indices of asteroids based on the results of observations with the SBG and Zeiss-1000 telescopes. The SBG telescope observations given only one color index V–R. The names of asteroids for which the color indices were determined for the first time are in bold type. If there are color indices estimated by other authors,

** ALCDEF, https://alcdef.org/

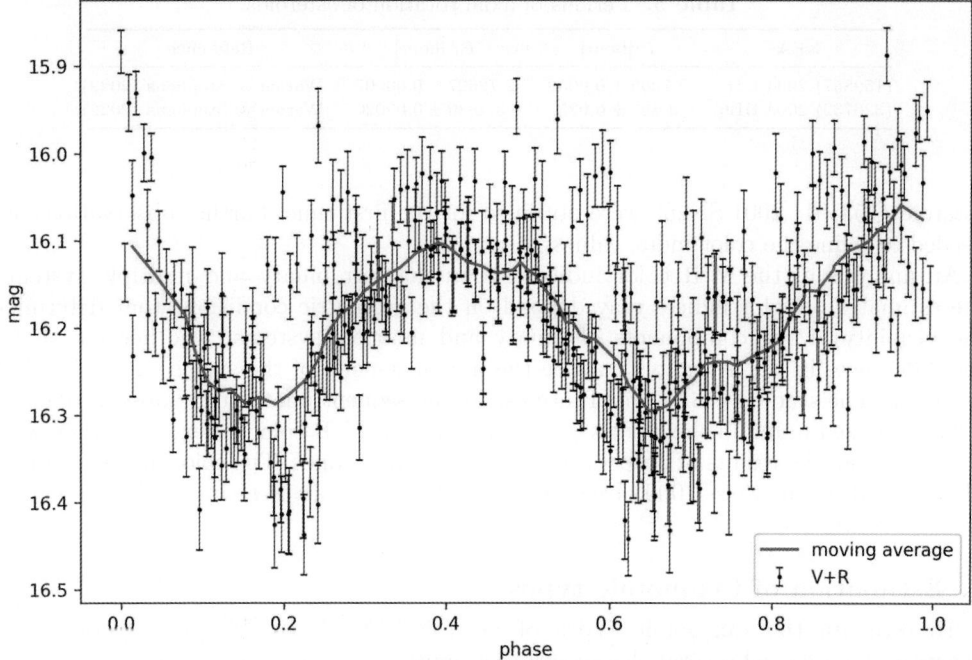

Figure 3. The phase light curve of the asteroid (159857) 2004 LJ1.

they are also given in Table 6. In most cases, the defined color indices match within the error ranges the values from the ALCDEF database. Let's consider individual cases.

The color index V–R $= 0.36^m$ for the asteroid (137170) 1999 HF1, determined from observations with the Zeiss-1000 telescope, is closer to the result V–R $= 0.37^m$ given Ieva *et al.* (2020) (Table 6). At the same time, observations with the SBG telescope give an estimate of V–R $= 0.41^m$, which is closer to the value of V–R $= 0.42^m$ (Polishook & Brosch (2008)) (Table 6). This may be due to the possible binary nature of the asteroid (137170) 1999 HF1 (Pravec *et al.* (2002)), when the binary system is observed in different configurations. However, recent results Warner (2016) do not confirm the presence of a companion for (137170) 1999 HF1. Further observations are needed to evaluate the V–R color index.

The situation is similar with estimates of the V–R color index for the asteroid (326732) 2003 HB6 (Table 6). The color index V–R $= 0.33^m$ obtained from observations with the Zeiss-1000 telescope is close to the estimate V–R $= 0.388^m$ (Lin *et al.* (2018)). The value V–R $= 0.43^m$ obtained with the SBG telescope agrees with V–R $= 0.429^m$ (Pravec *et al.* (2021)). The difference in the V–R estimates is probably due to the fact that the asteroid (326732) 2003 HB6 can be a binary (Warner & Stephens (2022)). Continued observations will resolve this uncertainty.

For asteroid (159857) 2004 LJ1, the V–R color indices obtained with the Zeiss-1000 and SBG telescopes are close to each other: V–R $= 0.46^m$ and 0.44^m (Table 6), but differ from the result V–R $= 0.58^m$ (Ieva *et al.* (2018)). It is possible that this difference is due to the fact that the asteroid (159857) 2004 LJ1 can be a binary (Warner & Stephens (2022)). Further photometric observations of the asteroid are required.

For the triple system (153591) 2001 SN263 (Nolan *et al.* (2008); Becker *et al.* (2015)) the V–R color indices obtained with the Zeiss-1000 and SBG telescopes (Table 6) differ by more than a standard deviation. Note that the estimates of the color indices of the

Table 5. Periods of axial rotation of asteroids.

NEA	P [hour]	P_0 [hour]	Reference
(159857) 2004 LJ1	2.7265 ± 0.0059	2.72627 ± 0.00007	Warner & Stephens (2022)
(326732) 2003 HB6	3.457 ± 0.035	3.4630 ± 0.0002	Warner & Stephens (2022)

asteroid (153591) 2001 SN263 were obtained for the first time. Further observations are needed to refine the color index values.

An analysis of the V–R color indices shows that for binary and multiple systems, the estimates of color indices may depend on the geometric conditions that determine the visibility of the components of binary and multiple systems. Another reason for the differences in the estimates may be the inhomogeneity of the surface of the studied asteroids. The studied NEAs are of interest for subsequent photometric observations.

We analyzed in detail the results of determining the V–R color index, since they were obtained from the results of observations with both telescopes. The color indices in other filters based on the Zeiss-1000 observations made it possible to estimate the taxonomic types of NEA.

5. Estimation of taxonomic types

To estimate the taxonomic types of asteroids based on multicolor photometric observations, we used several classification options.

(*a*) Tholen's classification based on B–V, V–R and V–I color indices (Dandy *et al.* (2015)) (Fig. 4).

(*b*) Bus–DeMeo classification based on B–R and V–I color indices (Ieva *et al.* (2018)) (Fig. 5a).

(*c*) Classification based on the average color indices B–R and V–I, obtained from the results of photometric observations of asteroids of known types (Ieva *et al.* (2018)) (Fig. 5b).

To determine the type of NEA, color indices from Table 6 were used.

For the asteroid (159857) 2004 LJ1, Tholen's classification (Fig. 4) indicates possible types C, S, or X. Classification by average color values (Ieva *et al.* (2018)) (Fig. 5b) indicates possible types D, S or X. According to the Bus–DeMeo classification (Fig. 5a), asteroid (159857) 2004 LJ1 can be classified as type S, which is consistent with the result of (Ieva *et al.* (2018)).

For the asteroid (326732) 2003 HB6 according to Tholen's classification (Fig. 4), taking into account that the B–V color index can reach 0.43^m (Table 6), types D or T are possible. The Bas–DeMeo classification and classification by average color indices (Ieva *et al.* (2018)) (Fig. 5) indicate the proximity of the asteroid (326732) 2003 HB6 to type D, which is consistent with the result of (Binzel *et al.* (2019)).

For the asteroid (153591) 2001 SN263, the classifications of Tholen (Fig. 4) and Bus–DeMeo (Fig. 5a) do not allow estimation of the taxonomic class. Classification by average color indices (Ieva *et al.* (2018)) (Fig. 5b) indicates closeness to type C, which is consistent with the results of Pajuelo *et al.* (2018) and Hasegawa *et al.* (2018).

6. Discussion and Conclusions

The results show that the observations on the SBG telescopes of the Kourovka Astronomical Observatory of the Ural Federal University and the Zeiss-1000 telescopes of the Simeiz Observatory of the INASAN complement each other and make it possible

Table 6. Asteroid color indices.

B–V	B–R	V–R	V–I	R–I	Reference
(137170) 1999 HF1					
$0.69^m \pm 0.02^m$	$1.05^m \pm 0.02^m$	$0.36^m \pm 0.02^m$			Zeiss-1000
		$0.41^m \pm 0.02^m$			SBG
$0.69^m \pm 0.03^m$		$0.42^m \pm 0.03^m$			Polishook & Brosch (2008)
$0.65^m \pm 0.08^m$		$0.37^m \pm 0.07^m$			Ieva et al. (2020)
(138127) 2000 EE14					
$0.61^m \pm 0.07^m$	$1.12^m \pm 0.07^m$	$0.48^m \pm 0.07^m$			Zeiss-1000
(153591) 2001 SN263					
$0.779^m \pm 0.010^m$	$1.054^m \pm 0.010^m$	$0.276^m \pm 0.008^m$	$0.527^m \pm 0.009^m$	$0.251^m \pm 0.009^m$	Zeiss-1000
		$0.39^m \pm 0.08^m$			SBG
(159857) 2004 LJ1					
$0.76^m \pm 0.11^m$	$1.25^m \pm 0.10^m$	$0.46^m \pm 0.10^m$	$0.82^m \pm 0.08^m$	$0.33^m \pm 0.04^m$	Zeiss-1000
		$0.44^m \pm 0.08^m$			SBG
$0.78^m \pm 0.05^m$		$0.58^m \pm 0.05^m$	$0.90^m \pm 0.05^m$		Ieva et al. (2018)
(326732) 2003 HB6					
$0.75^m \pm 0.05^m$	$0.98^m \pm 0.05^m$	$0.33^m \pm 0.08^m$	$0.89^m \pm 0.04^m$	$0.66^m \pm 0.04^m$	Zeiss-1000
		$0.43^m \pm 0.04^m$			SBG
$0.676^m \pm 0.124^m$		$0.388^m \pm 0.048^m$	$0.786^m \pm 0.054^m$		Lin et al. (2018)
		$0.429^m \pm 0.010^m$			Pravec et al. (2021)
(332446) 2008 AF4					
		$0.44^m \pm 0.08^m$			SBG
		$0.43^m \pm 0.02^m$			Hromakina et al. (2021)
2010 TV149					
$1.07^m \pm 0.08^m$	$1.42^m \pm 0.09^m$	$0.35^m \pm 0.09^m$			Zeiss-1000
2015 NU13					
		$0.40^m \pm 0.10^m$			SBG
		$0.388^m \pm 0.010^m$			ALCDEF

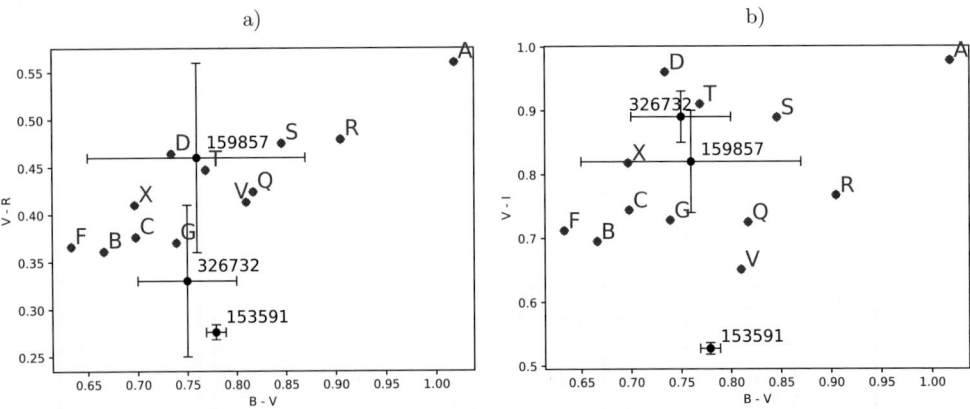

Figure 4. Color indices of the main types of asteroids according to Tholen's classification (Dandy *et al.* (2015)).

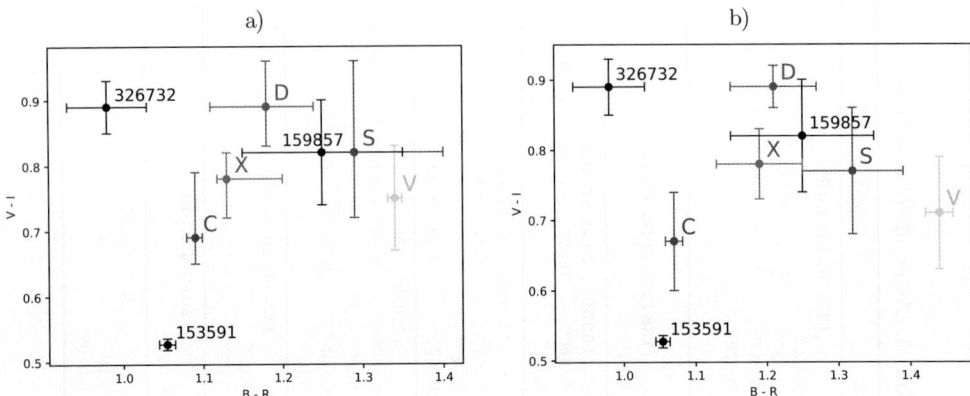

Figure 5. Color indices of the main types of asteroids: a) Bus–DeMeo classification (Ieva *et al.* (2018)), b) classification based on average color values (Ieva *et al.* (2018)).

to determine both the dynamic and physical characteristics of NEAs. The SBG telescope is focused on positional observations of NEAs, as well as on photometric ones in order to determine the rotation periods of NEAs and the color index V–R. Multicolor photometry on the Zeiss-1000 telescope provides reliable estimates of several color indices, on the basis of which the taxonomic class of NEA can be determined.

Improved orbital elements and reliable estimates of the A_2 acceleration due to the influence of the Yarkovsky effect makes it possible to conduct detailed studies of the orbital evolution of NEAs. Estimating the positive or negative value of the acceleration A_2 makes it possible to determine the direction of the NEA axial rotation, which, with a known period of axial rotation, makes it possible to analyze the geometric conditions of the asteroid's approach to the Earth.

Differences in the estimates of color indices, in particular V–R, obtained with different telescopes, as well as by different authors, indicate the relevance of multicolor NEA photometry. To compare the obtained color indices, it is necessary to analyze the geometric conditions of approach: change in the phase angle, the range of mean latitudes of the central point NEA's "disk", the location of satellites in binary and multiple systems, etc.

The solution of this problem will allow obtaining more reliable estimates of the taxonomic types of NEA.

7. Acknowledgments

The work was supported by the Ministry of Science and Higher Education of the Russian Federation, projects: FEUZ-2020-0030 (observations with the SBG telescope and their processing), FEUZ-2020-0038 (observations with the Zeiss-1000 telescope and their processing). The following telescopes were used during the observations: SBG, which is part of the scientific equipment shared research facility "Kourovka Astronomical Observatory" of UrFU, and Zeiss-1000, which is part of the scientific equipment of shared research facility "Terskol Observatory" of INASAN.

References

Becker, T.M., Howell, E.S., Nolan, M.C., Magri, C., Pravec, P., Taylor, P.A., Oey, J., Higgins, D., Világi, J., Kornoš, L., Galád, A., Gajdoš, Š., Gaftonyuk, N.M., Krugly, Y.N., Molotov, I.E., Hicks, M.D., Carbognani, A., Warner, B.D., Vachier, F., Marchis, F., & Pollock, J.T., 2015, *Icarus*, 248, 499

Binzel, R.P., DeMeo, F.E., Turtelboom, E.V., Bus, S.J., Tokunaga, A., Burbine, T.H., Lantz, C., Polishook, D., Carry, B., Morbidelli, A., Birlan, M., Vernazza, P., Burt, B.J., Moskovitz, N., Slivan, S.M., Thomas, C.A., Rivkin, A.S., Hicks, M.D., Dunn, T., Reddy, V., Sanchez, J.A., Granvik, M., & Kohout, T., 2019, *Icarus*, 324, 41

Dandy, C.L., Fitzsimmons, A., & Collander-Brown, S.J., 2003, *Icarus*, 163, 363

Galushina, T.Y., Bykova, L.E., Letner, O.N., & Baturin, A.P. 2019, *Astronomy and Computing*, 29, 100301

Hasegawa, S., Kuroda, D., Kitazato, K., Kasuga, T., Sekiguchi, T., Takato, N., Aoki, K., Arai, A., Choi, Y.-J., Fuse, T., Hanayama, H., Hattori, T., Hsiao, H.-Y., Kashikawa, N., Kawai, N., Kawakami, K., Kinoshita, D., Larson, S., Lin, C.-S., Miyasaka, S., Miura, N., Nagayama, S., Nagumo, Y., Nishihara, S., Ohba, Y., Ohta, K., Ohyama, Y., Okumura, S.-i., Sarugaku, Y., Shimizu, Y., Takagi, Y., Takahashi, J., Toda, H., Urakawa, S., Usui, F., Watanabe, M., Weissman, P., Yanagisawa, K., Yang, H., Yoshida, M., Yoshikawa, M., Ishiguro, M., & Abe, M., 2018, *Publications of the Astronomical Society of Japan*, 70, 114

Hromakina, T., Birlan, M., Barucci, M.A., Fulchignoni, M., Colas, F., Fornasier, S., Merlin, F., Sonka, A., Petrescu, E., Perna, D., Dotto, E., & Neorocks Team 2021, *Astron. and Astrophys.*, 656, A89

Ieva, S., Dotto, E., Mazzotta Epifani, E., Perna, D., Rossi, A., Barucci, M.A., Di Paola, A., Speziali, R., Micheli, M., Perozzi, E., Lazzarin, M., & Bertini, I. 2018, *Astron. and Astrophys.*, 615, A127

Ieva, S., Dotto, E., Mazzotta Epifani, E., Perna, D., Fanasca, C., Lazzarin, M., Bertini, I., Petropoulou, V., Rossi, A., Micheli, M., & Perozzi, E. 2020, *Astron. and Astrophys.*, 644, A23

Izmailov, I.S., Khovricheva, M.L., Khovrichev, M.Yu., Kiyaeva, O.V., Khrutskaya, E.V., Romanenko, L.G., Grosheva, E.A., Maslennikov, K.L., & Kalinichenko, O.A. 2010, *Astronomy Letters*, 36, 349

Krushinsky, V.V. 2018, in: E.D. Kuznetsov, D.S. Wiebe, A.B. Ostrovskii, S.V. Salii, A.M. Sobolev, K.V. Kholshevnikov, & B.M. Shustov (eds.), *Fizika Kosmosa: trudy 47-y Mezhdunarodnoy studencheskoy nauchnoy konferentsii (Yekaterinburg, 29 yanv. - 2 fevr. 2018 g.)* (Yekaterinburg: Ural University Press), p. 205

Lafler, J., & Kinman, T. D. 1965, *Astrophys. J. Supp.*, 11, 216

Lin, C.-H., Ip, W.-H., Lin, Z.-Y., Cheng, Y.-C., Lin, H.-W., & Chang, C.-K. 2018, *Plan. and Space Sci.*, 152, 116

Muinonen, K., Belskaya, I.N., Cellino, A., Delbò, M., Levasseur-Regourd, A.-C., Penttilä, A., & Tedesco, E.F. 2010, *Icarus*, 209, 542

Nolan, M.C., Howell, E.S., Benner, L.A.M., Ostro, S.J., Giorgini, J.D., Busch, M.W., Carter,

L.M., Anderson, R.F., Magri, C., Campbell, D.B., Margot, J.L., & Vervack, R. 2008, *Central Bureau Electronic Telegrams*, 1254

Oszkiewicz, D.A., Muinonen, K., Bowell, E., Trilling, D., Penttilä, A., Pieniluoma, T., Wasserman, L.H., & Enga, M.-T. 2011, *Journal of Quantitative Spectroscopy & Radiative Transfer*, 112, 1919

Pajuelo, M., Birlan, M., Carry, B., DeMeo, F.E., Binzel, R.P. & Berthier, J., 2018, *Monthly Not. Roy. Astron. Soc.*, 477, 5590

Penttilä, A., Shevchenko, V.G., Wilkman, O., & Muinonen, K. 2016, *Plan. and Space Sci.*, 123, 117

Polishook, D., & Brosch, N. 2008, *Icarus*, 194, 111

Pravec, P., Šarounová, L., Hicks, M.D., Rabinowitz, D.L., Wolf, M., Scheirich, P., & Krugly, Y.N. 2002, *Icarus*, 158, 276

Pravec, P., Hornoch, K., Kucakova, H., Kusnirak, P., Fatka, P., Jehin, E., Ferrais, M., Devogele, M., Kareta, T., & Moskovitz, N. 2021, *Central Bureau Electronic Telegrams*, 5039

Reddy, V., Kelley, M.S., Dotson, J., Farnocchia, D., Erasmus, N., Polishook, D., Masiero, J., Benner, L.A.M., Bauer, J., Alarcon, M.R., Balam, D., Bamberger, D., Bell, D., Barnardi, F., Bressi, T.H., Brozovic, M., Brucker, M.J., Buzzi, L., Cano, J., Cantillo, D., Cennamo, R., Chastel, S., Omarov, C., Choi, Y.-J., Christensen, E., Denneau, L., Dróżdż, M., Elenin, L., Erece, O., Faggioli, L., Falco, C., Glamazda, D., Graziani, F., Heinze, A.N., Holman, M.J., Ivanov, A., Jacques, C., van Rensburg, P.J., Kaiser, G., Kamiński, K., Kamińska, M.K., Kaplan, M., Kim, D.-H., Kim, M.-J., Kiss, C., Kokina, T., Kuznetsov, E., Larsen, J.A., Lee, H.-J., Lees, R.C., de León, J., Licandro, J., Mainzer, A., Marciniak, A., Marsset, M., Mastaler, R.A., Mathias, D.L., McMillan, R.S., Medeiros, H., Micheli, M., Mokhnatkin, A., Moon, H.-K., Morate, D., Naidu, S.P., Nastasi, A., Novichonok, A., Ogłoza, W., Pál, A., Pérez-Toledo, F., Perminov, A., Petrescu, E., Popescu, M., Read, M.T., Reichart, D.E., Reva, I., Roh, D.-G., Rumpf, C., Satpathy, A., Schmalz, S., Scotti, J.V., Serebryanskiy, A., Serra-Ricart, M., Sonbas, E., Szakáts, R., Taylor, P.A., Tonry, J.L., Tubbiolo, A.F., Veres, P., Wainscoat, R., Warner, E., Weiland, H.J., Wells, G., Weryk, R., Wheeler, L.F., Wiebe, Y., Yim, H.-S., Żejmo, M., Zhornichenko, A., Zoła, S., & Michel, P. 2022, *The Planetary Science Journal*, 3, 123

Warner, B.D. 2016, *Minor Planet Bulletin*, 43, 311

Warner, B.D., & Stephens, R.D. 2022, *Minor Planet Bulletin*, 49, 22

Astronomical Hazards for Life on Earth
Proceedings IAU Symposium No. 374, 2025
G. Tancredi, ed.
doi:10.1017/S174392132400070X

Asteroid Apophis and its associated fireballs

Gulchehra Kokhirova⊙ and Pulat Babadzhanov

Institute of Astrophysics, National Academy of Sciences of Tajikistan,
Postbus 734063, Dushanbe, Tajikistan
email: kokhirova2004@mail.ru

Abstract. The orbital evolution of NEA Apophis was investigated for the time interval of 12 kyrs, geocentric coordinates of radiants and velocities of theoretically related meteor showers were calculated, as well the dates of their activity were determined. As a result of a search among the fireball observations, on two fireballs for northern and southern branches of the nighttime shower were found. These fireballs, having the parameters close to predicted ones, probably, were produced by the fragments of the asteroid Apophis. It is recommended to carry out meteor and fireball observations on the predicted dates of shower activity of ±7 days in order to record meteor phenomena, possibly generated by fragments of asteroid Apophis.

Keywords. Meteors, meteoroids, minor planets, asteroids

1. Introduction

Near-Earth asteroid (NEA), 2004 MN4, was discovered on June 19, 2004 by observers R. A. Tucker, D. J. Tholen, and F. Bernardi at the Kitt Peak Observatory (USA) (Chesley 2006). Already in December 2004, preliminary calculations were made and the possibility of its collision with the Earth in 2029 was predicted. Additional observations and calculations carried out some time later ruled out this possibility, but revealed a chance of a collision in future. Thus, the NEA was classified as potentially dangerous for the Earth (Potentially Hazardous Asteroid), and on July 19, 2005, the asteroid was named Apophis (99942) in honor of the Egyptian god of evil and destruction, who lived in eternal darkness.

Apophis measures almost 340 m and the impact at a speed of 12.59 km s^{-1} with the Earth (http://neo.jpl.nasa.gov/jpl144 2022), which can occur as a result of close approach, will cause a catastrophe, the scale of which is difficult to foresee. According to various estimates, the energy released during such a collision will be equivalent to 800 MT in TNT equivalent. The consequences of such a catastrophe can be detrimental to life on the Earth. In addition, it is difficult to accurately assess the effect of such a very close encounter on the change in the asteroid's orbit.

2. Dynamical features of Apophis

Naturally, Apophis is a frequently observed object; more than 7000 observations are known, including radar and spectral ones, made over the period from 2004 to 2022. But its orbit is already so severely limited by existing observations that new astrometric data can only be used to improve the orbit, and therefore update the Risk Table data, only if it is particularly accurate. According to the elements of the orbit, Apophis belongs to the asteroids of the Aten group (see, e.g., Binzel *et al.* 2007, Giorgini *et al.* 2007, Rubincam 2007, and Sokolov *et al.* 2008). However, as a result of a reliably established approach to the Earth at a distance of 37–38 thousand km (accuracy of the order of a thousand

Table 1. Orbital elements of PHA Apophis (2000.0).

a AU	e	q AU	Q AU	i deg.	Ω deg.	ω deg.	π deg.	M deg.	Period d.
0.923	0.191	0.746	1.099	3.339	203.958	126.600	330.558	58.065	323.74

Table 2. Basic physical features of NEA Apophis.

H mag.	SMASSII spectral type	p	D km	Rotation period h
19.09	Sq	0.35	0.340	30.56

Table 3. Close approaches of Apophis with the Earth (from 1950 till 2100).

Date (yy/mm/dd)	Nominal distance, AU	Probability of close approach
1957/04/01	0.07540	1
1972/12/24	0.07924	1
1980/12/18	0.07213	1
1990/04/14	0.03293	1
1998/04/14	0.02439	1
2029/04/13	**0.00025**	1
2051/04/20	0.04146	0.2
2066/09/16	0.06956	0.2
2080/05/09	0.08693	0.3
2087/04/07	0.09581	0.1

km), on April 13, 2029, the asteroid will move from the Aten group to the Apollo group (Minor Planet Circ. 2005).

Orbital elements for Epoch 2459800.5 (2022-Aug-09.0) TDB and some physical features of Apophis (http://neo.jpl.nasa.gov 2022) are given in Tables 1 and 2, respectively. Table 1 contains usual orbital elements: a-semimajor axis, e - eccentricity, q, Q-perihelion and aphelion distances, Ω-longitude of descending node, ω-argument of perihelion, π - longitude of perihelion, M - mean anomaly. Note, according to Tisserand parameter $T_j = 6.46$ the orbit of Apophis is classified as typical asteroidal ones. Absolute magnitude, SMASSII spectral type, geometric albedo, diameter and rotation period are arranged in Table 2 (http://neo.jpl.nasa.gov 2022).

According to the results of calculations of the evolution of the orbit of Apophis, for the period from 1950 to 2100 there will be about 23 close encounters of the asteroid with the Earth (http://neo.jpl.nasa.gov 2022). Table 3 lists only those for which the nominal distance between the NEA and the Earth is less than 0.1 AU, the approach in 2029, when the distance will be 37–38 thousand km, is marked in bold.

3. Study of the evolution of the Apophis orbit and the search for related fireballs

Thus, calculations of the evolution of the orbit of Apophis showed the possibility of numerous very close encounters of the asteroid with the Earth. On the other hand, the age of this object is quite large and in the past it could collide with other near-Earth objects. As a result of such collisions, smaller fragments could be ejected from its surface. Therefore, it makes sense to look for meteors and fireballs generated by these fragments, or to identify meteor phenomena associated with Apophis. To do this, following the method of establishing a connection between the NEA and meteoroid streams (Babadzhanov *et al.* 2008, Babadzhanov *et al.* 2009, and Babadzhanov & Obrubov 1992),

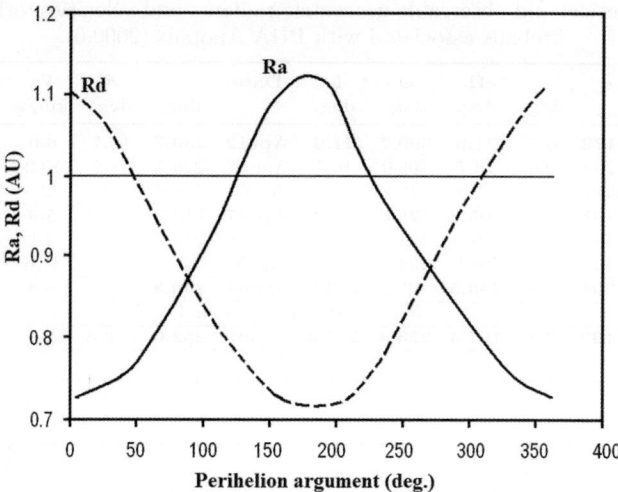

Figure 1. Dependences of the heliocentric distances of the ascending R_a and the descending R_d nodes of the Apophis orbit on the perihelion argument ω.

the Everhart method (Everhart 1974) was used to calculate the evolution of the asteroid orbit into the past over the period of one cycle of change in the perihelion argument.

Dependences of the heliocentric distances of the ascending R_a and the descending R_d nodes of the orbit on the perihelion argument ω and time T are shown in Figs. 1 and 2, respectively. The results of the study of evolution show that for one cycle of variation of the perihelion argument, R_a and R_d take values equal to 1 AU under four values of ω and, therefore, the condition of crossing the Earth's orbit is satisfied. If Apophis has a related meteoroid stream, then under these conditions the stream forms a nighttime and daytime meteor or fireball showers with northern and southern branches each. These four showers can be observed on the Earth at periods of time corresponding to the longitudes of the nodes of the orbit. The theoretical geocentric coordinates of radiants and the velocities of possible associated meteor showers were calculated from the elements of the NEA orbit at the found four values of the perihelion argument, and the longitudes of the Sun and the corresponding dates of activity of these showers were determined. According to the theoretical parameters in all published catalogs, an automatic search was performed for observed showers and individual meteors or fireballs with parameters close to those predicted. The results are presented in Table 4, where the theoretical showers are denoted by Latin letters A, B, C, D and highlighted in bold.

As can be seen from Table 4, four fireballs were found for the northern and southern branches of the night shower according to the observations of the US Prairie Network (PN) (McCrosky *et al.* 1978) and according to the data of the IAU Meteor Orbital Data Center (MODC) (https://www.ta3.sk/IAUC22DB/MDC2007 2022). Satisfactory agreement between the theoretical and observed parameters is confirmed by the values $D_{SH} \leq 0.1$, where D_{SH} is the Southworth and Hawkins criterion (Southworth & Hawkins 1963), which is a measure of the similarity of two orbits, in this case, a measure of the similarity of the theoretically predicted shower orbit and the observed fireball orbit, and the proximity of the values of the coordinates of radiants, velocities and periods of activity. As a result of a search, it was not possible to identify the predicted showers with any of the observable showers. This is associated with rather low geocentric velocities, which significantly reduce the possibility of detecting meteors and fireballs, and confirms that the object under study is a real asteroid.

Table 4. Theoretical and observable geocentric radiants and velocities, orbital elements of fireballs associated with PHA Apophis (2000.0).

Shower/ fireball	q AU	e	i deg.	Ω deg.	ω deg.	$L.$ deg.	Date	α deg.	δ deg.	V_g km/s	D_{SH}	Type	Catalogue
A	0.760	0.172	6.7	21.6	309.7	21.6	Apr12	230.7	15.7	6.0	-	N	-
2215	0.740	0.200	9.0	16.7	306.0	16.7	Apr 07	226.6	18.8	13.2	0.06	N	MODC
14806	0.695	0.211	6.0	17.0	318.7	17.0	Apr07	229.6	9.8	12.5	0.08	N	MODC
B	0.746	0.191	3.3	205.2	125.6	25.2	Apr17	215.2	-31.0	5.9	-	N	-
690407	0.809	0.130	1.7	197.8	134.4	17.8	Apr08	212.7	-27.4	3.1	0.09	N	PN
255	0.808	0.128	1.7	197.9	134.4	17.9	Apr08	216.7	-30.2	11.6	0.09	N	MODC
C	0.719	0.201	3.3	283.5	47.4	283.5	Jan03	259.8	-5.7	5.4	-	D	-
-							Meteors not found						
D	0.722	0.192	7.1	107.4	223.4	287.4	Jan07	252.0	-58.6	5.9	-	D	-
-							Meteors not found						

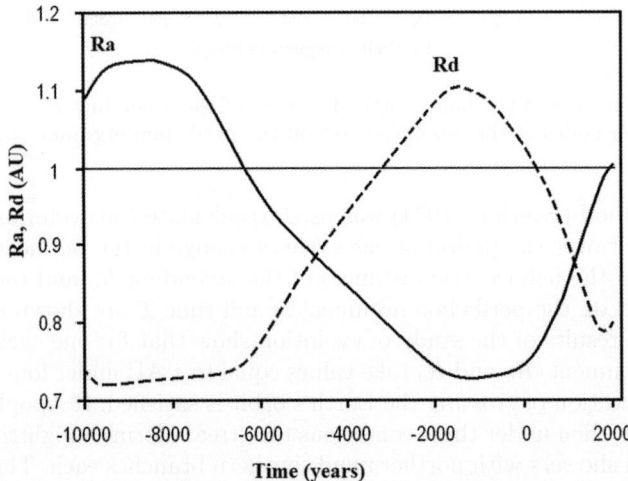

Figure 2. Dependences of the heliocentric distances of the ascending R_a and the descending R_d nodes of the Apophis orbit on time T.

4. Conclusion

Thus, as a result of the study, it is shown that the frequency of occurrence of crisis situations, when the orbit of Apophis is near the intersection with the Earth's orbit, is about 150 years (Table 3, Fig. 2) for each of the nodes, while the asteroid passes through all four nodes of the orbit in 6000 years. Therefore, if a close approach of the asteroid to the Earth does not occur during these 150 years, then the next series of such events can be expected in about 1500 years. Identified fireballs related to Apophis suggest that in the past there was at least one collision of an asteroid with another object, which led to the ejection of fragments from the surface of Apophis. Therefore, it is recommended to carry out fireball observations on predicted shower activity dates of ±7 nights in order to register new fireballs possibly related to asteroid Apophis.

References

Chesley, S.R. 2006, *Asteroids, Comets, Meteors: Proc. of the IAU Symp. 229, 2005, Cambridge Univ. Press*, 215

http://neo.jpl.nasa.gov/jpl144 2022

Binzel, R.P., Rivkin, A.S., Thomas, C.A., Vernazza, P., Burbine, T.H., DeMeo, F.E., Bus, S.J., Tokunaga, A.T., & Birlan, M. 2007, *Bulletin of the American Astronomical Society*, 39, 433

Giorgini, J.D., Benner, L.A., Ostro, S.J., Nolan, M.C., & Busch, M.W. 2007, *Bulletin of the American Astronomical Society*, 39, 512

Rubincam, D.P. 2007, *Icarus*, 192, 460

Sokolov, L.L., Bashakov, A.A., Pitjev, N.P. 2008, *Solar System Research*, 42, 18

2005 *Minor Planet Circ.*, 54567

http://neo.jpl.nasa.gov 2022

Babadzhanov, P.B., Williams, I.P., & Kokhirova, G.I. 2008, *Mon. Not. of the Royal Astron. Soc.*, 386, 1436

Babadzhanov, P.B., Williams, I.P., & Kokhirova, G.I. 2009, *Astron. and Astrophys.*, 507, 1067

Babadzhanov, P.B., & Obrubov, Yu.V. 1992, *Cel. Mech. and Dyn. Astron.*, 54, 111

Everhart, E. 1974, *Celestial Mechanics*, 10, 35

McCrosky, R. E., Shao, C.-Y., & Posen, A. 1978, *Meteoritika*, 37, 44, In Russian. NASA-supported research

https://www.ta3.sk/IAUC22DB/MDC2007 2022

Southworth, R.B. & Hawkins, G.S. 1963, *Smith. Contrib. Astrophys.*, 7, 261

Astronomical Hazards for Life on Earth
Proceedings IAU Symposium No. 374, 2025
G. Tancredi, ed.
doi:10.1017/S1743921324000747

On the mass indices of meteor bodies

Roman Zolotarev[ID] and Boris Shustov

Institute of Astronomy of the Russian Academy of Sciences, Pyatnitskaya Street, 48, Moscow, 119017, Russia

emails: rv_zolotarev@mail.ru, bshustov@inasan.ru

Abstract. A model of dynamical evolution of meteoroid swarm is applied to study the problem of difference in mass spectra of meteoric bodies during meteor showers and for sporadic meteors. It is demonstrated that mass spectra forms within meteoroid stream. Qualitative behavior of mass index in model is consistent with observational data.

Keywords. meteors, meteoroids, comets: general

1. Introduction

Properties of meteor bodies (meteoroids) is a subject of great interest in view of the problems of meteor showers, cometary activity and spacecraft protection. Formation of meteoroid streams, physical and dynamical evolution of the streams is a topic of active discussion. Distribution of meteoroids by mass can be described by

$$dN = C \cdot m^{-s} dm \tag{1}$$

where dN is number of particles in the mass range $[m, m + dm]$, s is mass index, C is normalizing factor. According to observational data (optical and radio-data), for meteoroids with masses in range 10^{-6} g $< m < 10^{-1}$ g, mass index $s \simeq 2$ for sporadic meteors and $s < 2$ near maximums of meteor showers, while at the edge of the stream s may exceed 2 (See detailed review in Shustov & Zolotarev (2022)).

The most direct way to deal with this problem is modeling of evolution of meteoroid streams with particles of different masses. We consider comets 96P/Machholz and 2P/Encke as examples of parent bodies. The main goal is to study variations of in relative abundance of particles of different masses in the stream over time. We modeled in two steps. First, we construct the model of ejection of particles from the parent comet, and then we model dynamical evolution of meteoroid swarm in Solar System and calculate meteoroid mass spectra in Earth vicinity. The result of solving the first problem is the velocity field of particles of various masses ejected from comet nucleus. This provides the necessary initial condition for solving the second problem which describes evolution of meteoroid swarm. A series of works have been dedicated to the study of the ejection process of particles from comet, a comprehensive review of models can be found in Ryabova (2013). General approach to study evolution of meteoroid streams is described in Ryabova (2020). In this paper we focus on detail on the model of meteoroid streams produced by comets 96P/Machholz and 2P/Encke. The problem statement and calculation method are described in Section 2, and results in Section 3.

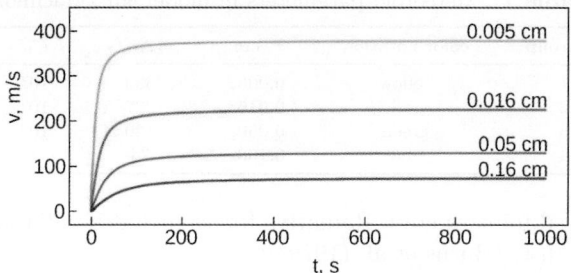

Figure 1. Time dependence of the escaping particle velocity v on time for particles of different radii in model 96P/Machholz.

2. Model

Mass index of sporadic meteor bodies is close to $s = 2$. Such mass spectra is typical for a wide variety of ensembles of astronomical objects. An almost "philosophical" explanation is following. Formation of ensembles of astronomical objects, as a rule, is a complex process and depends on many factors and competing processes. If there are many processes and all of them make a comparable contribution, their combined result can be represented as the effect of some random process, which can be described as statistical noise. With respect to mass spectra, such noise can be described by a power-law distribution with the slope exponent in the mass spectrum $s = 2$, which is typical for noise-type processes (Shustov (2019)). For example, the mass index of small dust grains found in comet 67P/Churyumov-Gerasimenko is close to $s = 2$ (Marschall et al. (2020)). Thus, we assume initial mass index $s_0 = 2$ for particles, escaping from nuclei of parent body.

Ejection model considers particles to be accelerated by gas outflow during a comet's perihelion passage. Particles are typically ejected from insolated semi-sphere of comet nucleus or in some solid angle up to the full sphere. Equation of motion for particle, escaping parent body is:

$$\frac{dv}{dt} = \frac{F_p}{m} - \frac{GM_c}{d^2} \tag{2}$$

where v and m are velocity and mass of the particle, F_p is outflow gas pressure force, M_c is comet core mass, d – distance to the core center. $F_p \sim \dot{M_c} (u_0 - v) r^2/d^2$, where u_0 is gas outflow velocity, r is particle radii (see Shustov & Zolotarev (2022) for more details). It can be seen that the acceleration decreases rapidly with distance from the comet nucleus. Figure 1 demonstrates time dependence of velocity of particle w.r.t. comet nuclei in model of 96P/Machholz.

After few dozens of seconds, the particle velocity stops changing and can be considered as terminal. It corresponds a distance about 10 nucleus radii. Dependence of terminal velocity on particle size can be approximated as $v_t \propto r^{-1/2}$.

To construct mass index it is necessary to model motion of particles with different masses. In our model we use particles of 4 different masses (sizes) with density $\rho = 2$ g/cm^3. Table 1 summarizes parameters of the particles in model 96P/Machholz. $N = 10^5$ particles of each size were considered in our dynamical model. C_n is normalizing factor (see below).

When the particles are sufficiently far away from the nucleus (at distance $R_{esc} \sim 10 \, R_{nucleus}$) their further motion can be considered as motion governed by general gravitational field of the Solar System taking into account the radiative forces. Model of the Solar System gravitational field includes gravity of the Sun and planets. The initial conditions for gravitating bodies and parent body were taken from HORIZONS

Table 1. Meteoroid parameters in model 96P/Machholz.

# group	color notation	r, cm	$v_t, m/s$	C_n
1	yellow	0.005,	333	10^9
2	red	0.016,	227	10^6
3	green	0.050,	130	10^3
4	blue	0.160,	74	1

database (https://ssd.jpl.nasa.gov). Radiative forces includes radiation pressure and the Poyting-Robertson drag (Burns et al. (1979)):

$$\mathbf{F} = \left(\frac{SAQ}{c}\right)\left[\left(1 - \frac{\dot{R}}{c}\right)\frac{\mathbf{R}}{R} - \frac{\mathbf{V}}{c}\right] \tag{3}$$

where S is the solar radiation flux at the distance of the particle, A is the geometrical cross section of the particle, Q is the scattering coefficient (in our model $Q = 1$), $\mathbf{R}$ and $\mathbf{V}$ are the position and velocity vectors of the particle relative to the Sun. Dynamics of particles was integrated with the REBOUND code using hybrid scheme MERCURIUS (Rein et al. (2019)).

After the simulation has done, meteoroid stream structure can be studied. In order to describe the spatial structure of the stream (and calculate mass index), a sufficient number of considered particles of each size is required. If the calculation is carried out directly, the number of the smallest particles should be many orders of magnitude larger, which is typically too large for calculations. For this reason, we simulate the motion of particles in all groups with the same number of particles. To obtain model of spatial density distribution of particles of each size, the resulting spatial density of these particles is then multiplied by the normalizing factor C_n (see Table 1). Number of particles $N = 10^5$ is sufficient to reproduce mass spectra in any part of the stream orbit.

For comets an ejection of particles occurs every time they pass the perihelion zone, but the intensity (mass of the ejected material) should gradually decrease. In our model it is assumed that the number of particles decreases exponentially with each subsequent ejection, i.e. number of particles on the k-th passage of the perihelion is $N_k = N f^k$, where $f < 1$ is a reducing factor.

Mass index values for a stream was calculated at a distance of 1 AU from the Sun. Fig. 2 illustrates a scheme for mass index calculation. The number of particles of each size is counted in the near-Earth sphere of radius R_a (counting sphere). Successive positions of the sphere corresponds to a change in the solar longitude with a step of 1°. Detailed description of the model can be found in Zolotarev & Shustov (2022).

3. Results

Evolution of meteoroid swarm associated with comet 96P/Machholz is shown in figure 3. Particles with different sizes is drawn with different colors (see Table 1). At the very beginning of evolution, when the ejection of particles has already occurred, but the parent body has passed through orbit only for a small part of the orbital period (first picture on Fig. 3), an extended trail of particles is formed, and the separation of the trajectories of particles of different sizes is clearly visible. Naturally, non-gravitational factors have the strongest effect on smaller particles. The most massive particles (shown by blue in Fig. 3) move away from the comet nucleus symmetrically at the maximum distance determined by the velocity of initial ejection. Lighter particles travel much further from the nucleus. It can be seen, that the smallest particles have already "forgotten" the initial ejection velocity and there is no symmetry at all. In general, the meteoroid ensemble forms an elongated structure, which over time becomes similar to a narrow

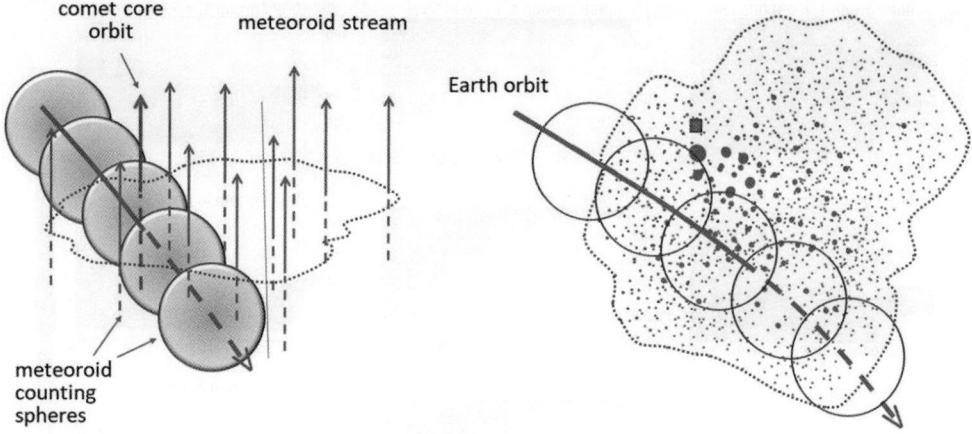

Figure 2. Schematic illustration of mass index evaluation.

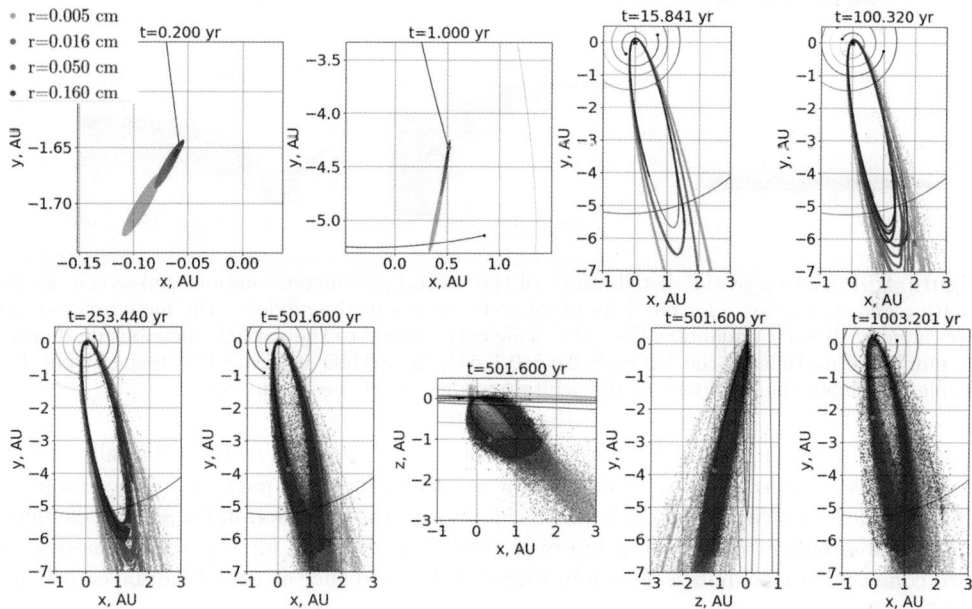

Figure 3. Evolution of the meteoroid stream in the model of stream associated with 96P/Machholz. XY-plane corresponds the plane of ecliptic. The color represents particles of different size (see notation in Table 1), initial number of particles in each group $N = 10^5$. Black dots and lines show the planets and their orbits, purple dot is the position of parent body at the specified time.

tail extended along the comet's orbit in both directions. The lagging part of the tail is longer than the front part. Such structures was found in observations and called dust trails (Jenniskens (2008)). One can notice various structural changes in the stream at later stages caused by gravitational factors, mainly, encounters with Jupiter. Additional branches in the stream appear.

Note that not a continuous size distribution of particles was considered and individual filaments are visible in Fig. 3. In fact, the particle size distribution in real meteoroid streams is continuous and the overall view of the stream in reality expected to be more blurry (uninterrupted).

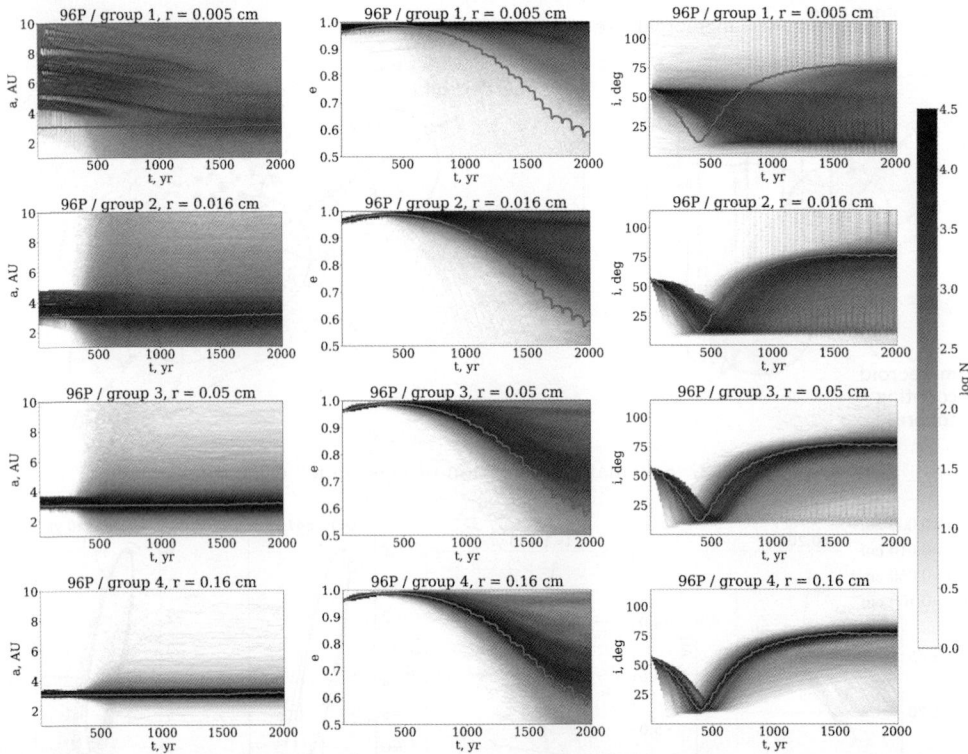

Figure 4. Evolution of the distributions of the orbital parameters: major semi-axis a, eccentricity e and inclination i for particles of different groups in the model of the meteoroid stream associated with 96P/Machholz. The color scale corresponds to the particle distribution density, i.e. number of particles in the intervals $\Delta a = 0.1$ AU, $\Delta e = 0.005$ and $\Delta i = 0.7°$ respectively. The purple line shows the evolution of the orbital parameters of the parent comet.

To estimate secular evolution of the meteoroid stream we use a method that allows us to track the evolution of the distribution of the orbital elements of particles in the ensemble. This includes constructing of diagram that represent the variation in the orbital elements of the meteoroids of each size group over time. The diagram for the stream associated with comet 96P/Machholz is shown in Figure 4. Vertical slice at time t can be considered as histogram of distribution of a given orbital element at the time t.

As it was mentioned above, the lighter particles quickly "forget" the original orbit of parent body and scatter in phase space. Heavier particles are less exposed to action of non-gravitational forces and "follow" the orbit of the parent body.

Let us consider the distribution of meteoroids at a distance of 1 AU in the ecliptic plane, in the region where the Earth crosses the densest part of the meteoroid stream. This corresponds to meteor shower event. Figure 5 (left) shows the distributions of stream particles (after normalization) in the counting spheres with $R_a = 0.1$ AU in 20-day period for model of stream associated with comet 96P/Machholz. Numbers of particles were calculated for the points corresponding to changes in the solar longitude with a step of 1 degree (i.e. almost every day). Age of the stream (time in the simulation from start) is about 500 yr. Reducing factor for repeated ejections was chosen $f = 0.97$. Note that this is the (normalized) number of the model particles, which allows to study relative characteristics of the stream. Mass index value s is one of these relative characteristics.

Right panel of the Fig. 5 shows variations of mass index corresponding to the data on the left panel. For comparison with the observations, the profile is plotted over the

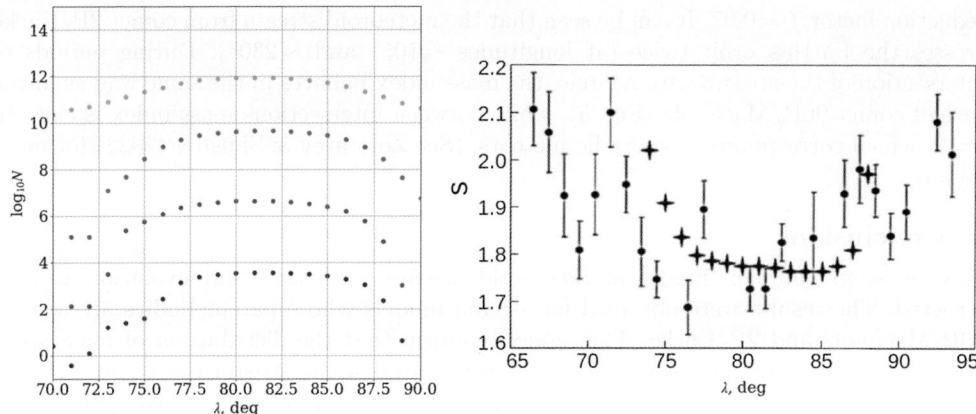

Figure 5. Distributions of number of model (normalized) particles in the model of stream associated with comet 96P/Machoolz (left) and corresponding mass index (right). On the right panel crosses is a model values, points corresponds the data on the Arietids stream (Blaauw et al. (2011)).

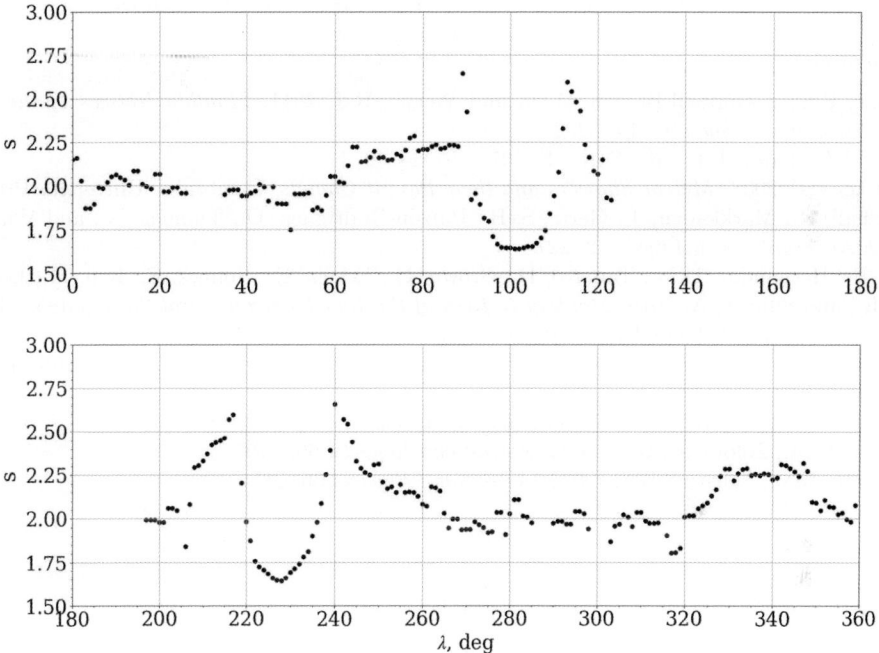

Figure 6. Mass index calculated for model of meteoroid stream associated with comet 2P/Encke in the region near Earth's orbit. Age of the stream is 511.6 yr.

observational data by Blaauw et al. (Blaauw et al. (2011)), which presents results of observations of the Arietids stream with the CMOR radar. The behavior of mass index s as the Earth passes through the stream is qualitatively reproduced in the model. At the maximum of the meteor stream $s < 2$, while at the edge of the stream $s \simeq$ or higher.

Figure 6 shows the distribution of the mass index in the model stream associated with comet 2P/Encke. The counting spheres are distributed with a step of 1 degree along a circle with a radius of 1 AU from the Sun in the ecliptic plane over the entire range of longitudes. Age of the stream also is about 500 years from the beginning of calculations, reduction factor $f = 0.97$. It can be seen that the meteoroid stream from comet 2P/Encke crosses the Earth's orbit twice (at longitudes $\sim 105°$ and $\sim 230°$). During periods of intersection of the stream with a circle, the mass index behaves in the same way as in the case of comet 96P/Machholz (Fig. 5), while between intersections mass index is close to $s = 2$, which corresponds to sporadic meteors. (See Zolotarev & Shustov (2022) for more details).

4. Conclusions

A two-step complex model of meteoroid stream formation and evolution is constructed. The results were obtained for stream models whose parent bodies are comets 96P/Machholz and 2P/Encke. The models confirm that the distribution of mass spectra is formed in the meteoroid stream: at the center of the stream the values of mass index $s < 2$, while at the edge of the stream s may exceed 2. There are two factors leads to changing in the structure of meteoroid stream. First, the initial velocity of particle ejection from the parent comet depends on the particle size and small particles move away from the nuclei faster, and second, small particles are more exposed to the action of radiative forces and scatter in space faster than large particles. Thus the value of the mass index at the center of the stream decreases. These results are consistent with the variations in mass index found from the observations of meteor streams.

References

Blaauw, R.C., Campbell-Brown, M.D., and Weryk, R.J. 2011, *Monthly Notices of the Royal Astronomical Society* 414, 3322

Burns, J.A., Lamy, P.L., and Soter, S. 1979, *Icarus* 40, 1

Jenniskens, P. 2008, *Meteor Showers and their Parent Comets*, Cambridge University Press

Marschall, R., Markkanen, J., Gerig, S.-B., Pinzón-Rodríguez, O., Thomas, N., and Wu, J.-S. 2020, *Frontiers in Physics* 8, 227

Rein, H., Hernandez, D.M., Tamayo, D., Brown, G., Eckels, E., Holmes, E., Lau, M., Leblanc, R., and Silburt, A. 2019, *Monthly Notices of the Royal Astronomical Society* 485, 5490

Ryabova, G.O. 2013, *Solar System Research* 47, 219

Ryabova, G.O. 2020, *Mathematical Modelling of Meteoroid Streams*, Springer International Publishing

Shustov, B.M. 2019, *INASAN Science Reports* 4, 356

Shustov, B.M., Zolotarev, R.V. 2022, *Astronomy Reports*, 66, 179

Zolotarev, R.V., Shustov, B.M. 2022, *Astronomy Reports* 66, 255

Astronomical Hazards for Life on Earth
Proceedings IAU Symposium No. 374, 2025
G. Tancredi, ed.
doi:10.1017/S174392132400098X

IAU374: Lunar Impact Events by Sharjah Lunar Impact Observatory (SLIO) in 2020

Ammar E. M. Abdulla[1,2], Mohammad F. Talafha[2], Mashhoor A. Al-Wardat[1,2] and Hamid. M. Al-Naimiy[1,2]

[1]Department of Applied Physics and Astronomy, College of Sciences, University of Sharjah, Sharjah 27272, UAE
email: amabdulla@sharjah.ac.ae

[2]Sharjah Academy for Astronomy, Space Sciences and Technology, University of Sharjah, Sharjah 27272, UAE

Abstract. Sharjah Lunar Impact Observatory (SLIO), established in 2020, is known to be the only observatory in the Middle East and North Africa region that fully focuses on lunar impact observation and analysis. The Observatory is located inside the cosmic garden of the Sharjah Academy for Astronomy, Space Sciences and Technology, University of Sharjah, Sharjah, United Arab Emirates. The coordinates of the site are: 25°17'02.1"N 55°27'48.4"E, with an altitude of 80 m above sea level. We present 5 lunar impact events that were detected by SLIO in the year 2020. The associated properties of the events were deduced from comprehensive analysis in which we have recorded apparent magnitudes of 7.94, 8.92, 9.54, 10.06, and 7.79 and associated durations of 0.04, 0.08, 0.08, 0.04, 0.08 seconds respectively. Essentially, since the Moon is the closest companion to our Earth, these meteorites represent possible dangers on Earth as well as on the Moon. Therefore, a continuous monitoring system that provides an estimation of number, size and distribution of meteorites hitting the lunar surface can allow to predict threats to Earth as it would give information about the meteorite activity in Earth's neighborhood, which can considerably help prevent potential disasters.

Keywords. SLIO; Sharjah; Meteorites; 2020; Dangers; Earth.

1. Introduction

The Sharjah Lunar Impact Observatory is an initiative by the Sharjah Academy of Astronomy, Space Sciences and Technology to embody the vision of establishing the entity as a leading institute in Space Science research in the Middle East and North Africa (MENA) region. The entity, which is a scientific institute belonging to the University of Sharjah, does its best to enhance research capabilities and academic programs as well as taking part in space exploration programs. The main goal of the establishment of the Sharjah Lunar Impact Observatory was to study meteorites with masses varying between grams and kilograms entering the Moon's thin atmosphere and making their way to the surface with a contact that produces a characteristic crater and an instantaneous burst of light (Ait Moulay Larbi et al. 2013; Baratoux et al., 2012; Talafha et al. 2020). This burst of light can be observed from Earth if it was bright enough, which is essentially what we attempt to detect at the Sharjah Lunar Impact Observatory and hence we aim to observe a sufficient area of the darker side of Moon which is usually about 8–10 days each month (Talafha et al. 2022). Lunar impacts are a valid representation of dangers on the Moon and the study of them is key, especially that they pose threats to Moon exploration projects as that is the tendency of a large number of major entities

Figure 1. Sharjah Lunar Impact (red rectangle) and Sharjah Optical Observatory (yellow rectangle) (Google Earth).

that make up the space sector not only nationally, but also globally (Avdellidou 2018; Cook, n.d.; Mohon 2021). Therefore, it is necessary to have estimations of the frequency, scale distribution, velocity, mass and size ranges of lunar impacts in order to predict threats to not only the Moon but also Earth, as the Moon is Earth's closest neighbor which means these two celestial objects more or less share the same meteor environment (Madiedo et al. 2015). The choices of observational instruments and associated softwares in Sharjah Lunar Impact Observatory were almost entirely inspired by NASA's Marshall Space Flight Center project which goal was to continuously monitor the darker side of the Moon to detect lunar impact events (Mohon 2021). In this paper, we briefly describe the role of the Sharjah Lunar Impact Observatory in observing and analyzing lunar impact events during 2020, the year of the observatory's establishment.

2. Sharjah Lunar Impact Observatory (SLIO)

The observatory is located in the campus of the University of Sharjah, precisely in the cosmic garden of Sharjah Academy of Astronomy, Space Sciences and Technology. The site has an altitude of 80 meters above sea level and its coordinates are 25° 17' 02.1" N 55° 27' 48.4" E. Figure 1 displays a satellite view of the site using Google Earth; Sharjah Lunar Impact Observatory is highlighted with a red rectangle, where the building highlighted by a yellow rectangle is the Sharjah Optical Observatory. This view can be accessed via the link: https://earth.google.com/web/search/ Sharjah+Academy+of+Astronomy,+Space+sciences+%26+Technology+-+Sheikh+ Khalifa+Bin+Zayed+Al+Nahyan+Road+-+Sharjah/@25.28461097,55.46115217, 18.98540353a,1016.78719193d,35y,-0h,0t,0r/data=CigiJgokCV59-agygTxAEd54Y2siyzV AGTLMyqguxE5AIYz6Q6ymDEhA

2.1. *Telescope*

Sharjah Lunar Impact Observatory uses a 14" LX 200 ACF Schmidt-Cassegrain tele-scope with a focal ratio of f/10 as shown in Figure 2 (dropped to f/3.3 by using a focal

Figure 2. The 14" LX 200 ACF Schmidt-Cassegrain telescope used at the Sharjah Lunar Impact Observatory.

Figure 3. 902H Ultimate CCD camera used at the Sharjah Lunar Impact Observatory (Watec, n.d).

reducer for a larger FoV of the darker side of the Moon). This telescope has a corrector plate in which light enters in parallel lines that are then reflected by a spherical primary mirror onto a secondary mirror that finally redirects it onto the eyepiece (Meade n.d.). The mount used is an alt-azimuth inside the observatory's dome.

2.2. *Camera*

A lunar impact event usually lasts for half a second at most, therefore, we need a continuous view of the darker side of the Moon for almost the whole duration of observation. For that, we attach a high-sensitivity video camera of the model 902H Ultimate shown in Figure 3 to the telescope. The camera is manufactured by Watec, has a 0.5" monochrome CCD ICX439ALL sensor and records at 25 frames per second providing a resolution of 720 x 576 pixels (Watec, n.d).

2.3. *Dome*

The telescope and camera assembly are hosted by a 9 feet (2.7 meters) high dome shown in Figure 4. The dome is manufactured by NexDome and it is protected by a thick roof with a manually-adjustable opening to access the sky.

Figure 4. The Sharjah Lunar Impact Observatory dome (left). Sharjah Academy of Astronomy, Space Sciences and Technology building (right).

3. Observation

The lunar impact observation process in Sharjah Lunar Impact Observatory is summarized in 4 steps as such:

(*a*) Attach the focal reducer and the CCD camera to the telescope then connect the camera to a computer via an EasierCAP analogue-to-digital converter.

(*b*) Point the telescope to the dark side of the Moon and start recording via the capture software SharpCap in 30 second intervals.

(*c*) Manually adjust the telescope using the view provided by SharpCap software to remain directed at the darker (This can be done automatically using the Autostar handset, but the handset does not account for the attachment of the focal reducer).

(*d*) Compile the video files and prepare them for analysis.

The final observational setup is shown in Figure 5, where the focal reducer is connected to the 14" telescope followed by the CCD camera.

It is important to note that the number of observational sessions/intervals per night highly depends on sky clarity and the interference of the brighter side of the Moon, therefore, it is difficult to have a fixed number of AVI files for every observation night.

4. Analysis

The analysis process of the AVI files obtained from observation is summarized in 5 steps as such:

(*a*) Open the analysis software LunarScan 1.5 and define the area to be scanned using a raw clip of the Moon, an example is shown in Figure 6.

(*b*) Add a dark frame as an AVI digitized dark field for each scanning session.

(*c*) Plug in a set of AVI files and run the scanning process after adjusting for the proper scanning settings (these are dependent on observational conditions, clarity/quality of captured files, preference, etc).

(*d*) View and closely study all the detections provided by the software: confirm for possibly true events (candidates) and eliminate false events.

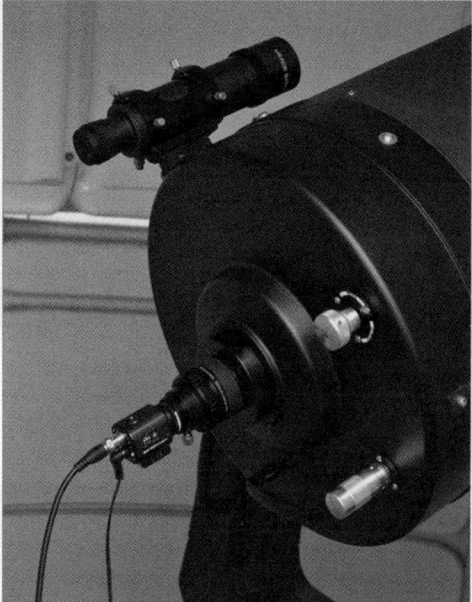

Figure 5. The observational setup at the Sharjah Lunar Impact Observatory (CCD camera − > f/3.3 focal reducer − > 14" telescope).

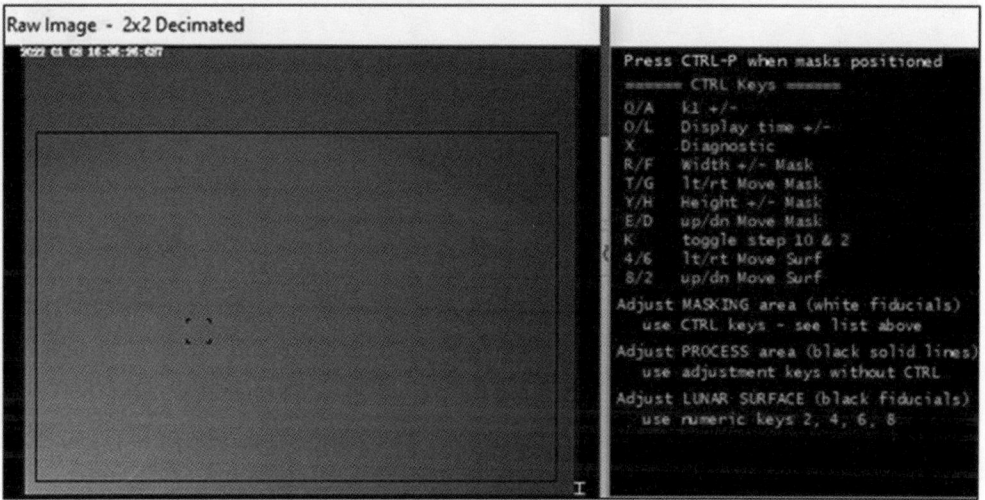

Figure 6. The interface of LunarScan 1.5 in the midst of fitting the masking area of the darker side of the Moon for scanning (left). The command window to use in order to move the masking area (right).

(*e*) Re-examine all the candidates further to eventually confirm or reject a lunar impact event. This involves a sophisticated inspection of the raw AVI files, the split frames and pixel distribution prior, during and after the flash.

5. Events Observed in 2020

5.1. *Event 1*

This lunar impact event was detected during the very first week of operation for the Sharjah Lunar Impact Observatory on the 1st of March 2020 at 16:53:22.969 UT with an

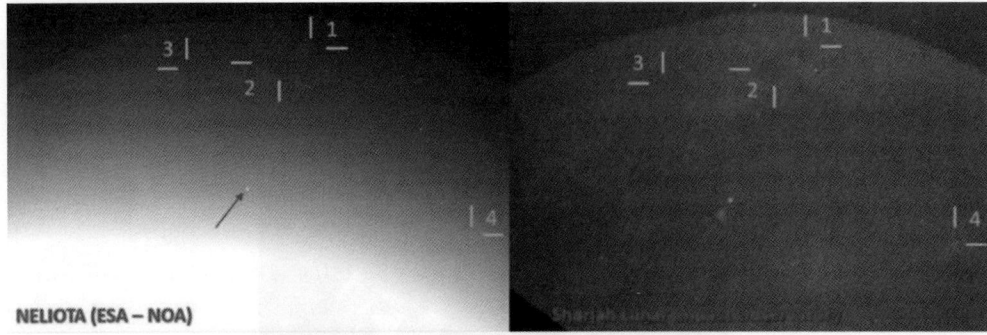

Figure 7. Event 1: Lunar impact event on the 1st of March 2020 at 16:53:22.969 UT with numbers representing the same features as viewed from NELIOTA (left) Sharjah Lunar Impact Observatory (right) (NELIOTA, n.d.; Talafha et al. 2022).

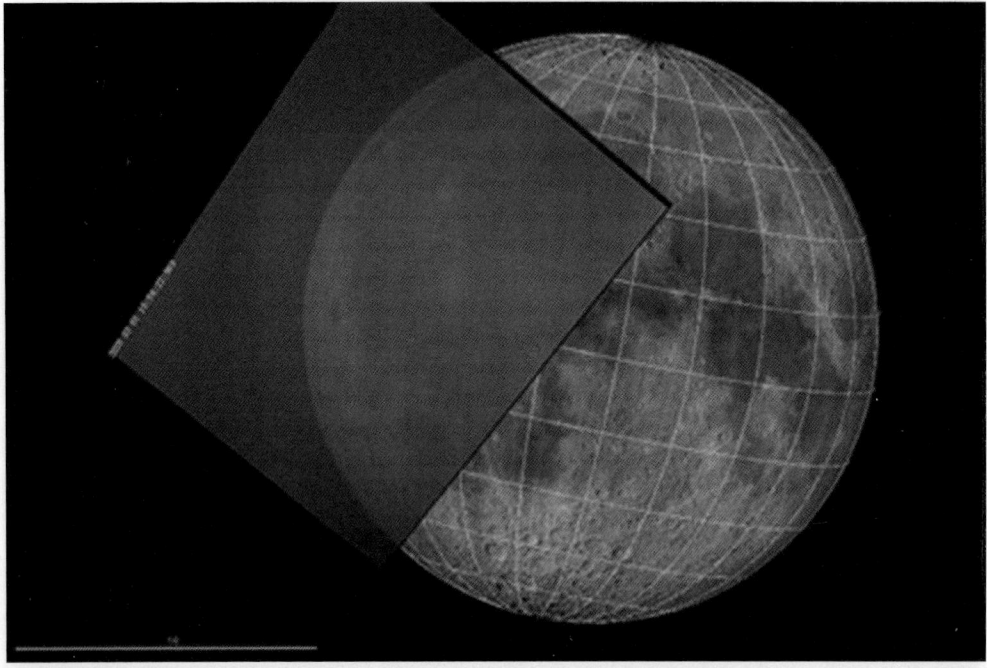

Figure 8. Event 1:Selenographic map created by the Sharjah Lunar Impact Observatory team to identify which part of the Moon was the lunar impact event on the 1st of March 2020 in.

apparent magnitude of 7.94, mass of 1.92 kg, diameter of 2.84 meters and belongs to the Delta Leonid Complex meteor shower. As well as the detection by Sharjah Lunar Impact Observatory in United Arab Emirates, this event was confirmed by the NELIOTA - ESA project in Greece. This double detection provided massive credibility to the Sharjah Lunar Impact Observatory considering the detection was only in the first week of operation! Figure 7 shows the exact location of the impact as viewed from both NELIOTA and Sharjah Lunar Impact Observatory, the numbers represent identical features of the Moon for demonstration purposes. Figure 8 shows which part of the Moon was affected by this lunar impact event in a selenographic map and Figure 9 displays how the event evolves with time in split frames provided by the software (Talafha. M n.d.).

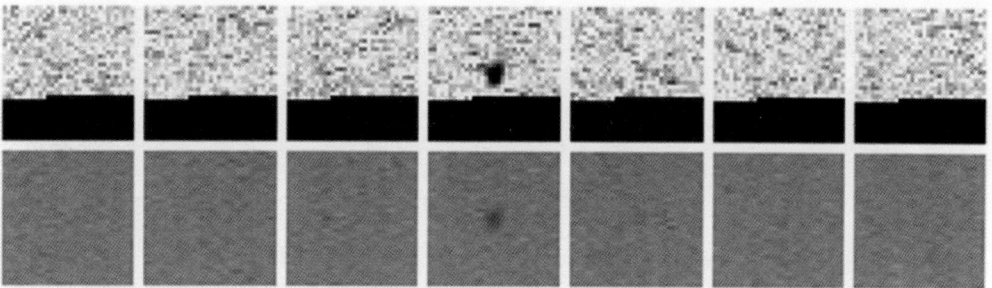

Figure 9. Event 1: Chronological frame sequence of the lunar impact event on the 1st of March 2020 provided by the software LunarScan 1.5 (Talafha et al. 2022).

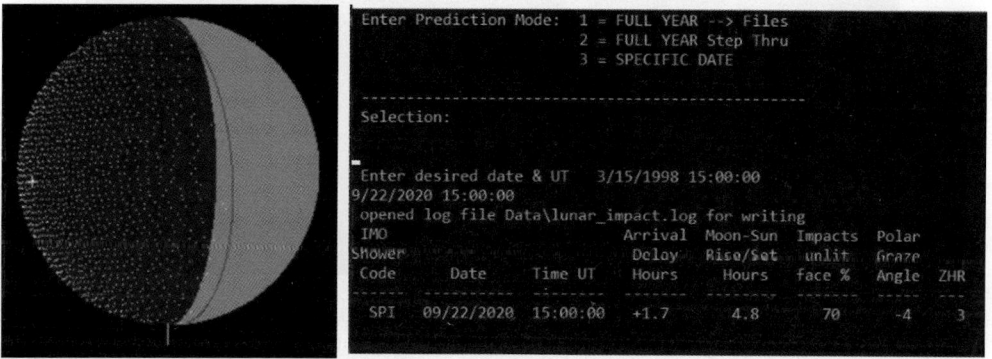

Figure 10. Event 2: The LunarScan 1.5 window of the lunar impact event on the 22nd of September 2020 where it can be seen on the left where the software thinks the event took place on the Moon (the plus sign).

5.2. *Event 2*

This lunar impact event was detected on the 22nd of September 2020 with a duration of 0.08 seconds, apparent magnitude of 7.94, mass of 0.24 kg, diameter of 0.09 meters and belongs to the Southern Delta Piscids meteor shower. This impactor is considerably smaller and less massive than Event 1 but larger and more massive than Events 3, 4 and 5. Event 1 and 2 having the same apparent magnitude even though the impactor in Event 2 is smaller and less massive could do with the velocity in which the impactor possessed prior to the contact. Figure 10 shows the display window of the software for Event 2 and Figure 11 shows a real-life capture of the brightest point of the flash.

5.3. *Event 3*

This lunar impact event was the first of two detected on the 20th of November 2020 with a duration of 0.08 seconds, apparent magnitude of 9.54, mass of 0.06 kg, diameter of 0.06 meters and belongs to the Alpha Monocerotids meteor shower. Figure 12 shows the display window of the software for Event 3.

5.4. *Event 4*

The second lunar impact event detected on the 20th of November 2020 was slightly shorter, fainter, less massive and smaller than the first one. It had a duration of 0.04 seconds, apparent magnitude of 10.06, mass of 0.017 kg and diameter of 0.04 meters.

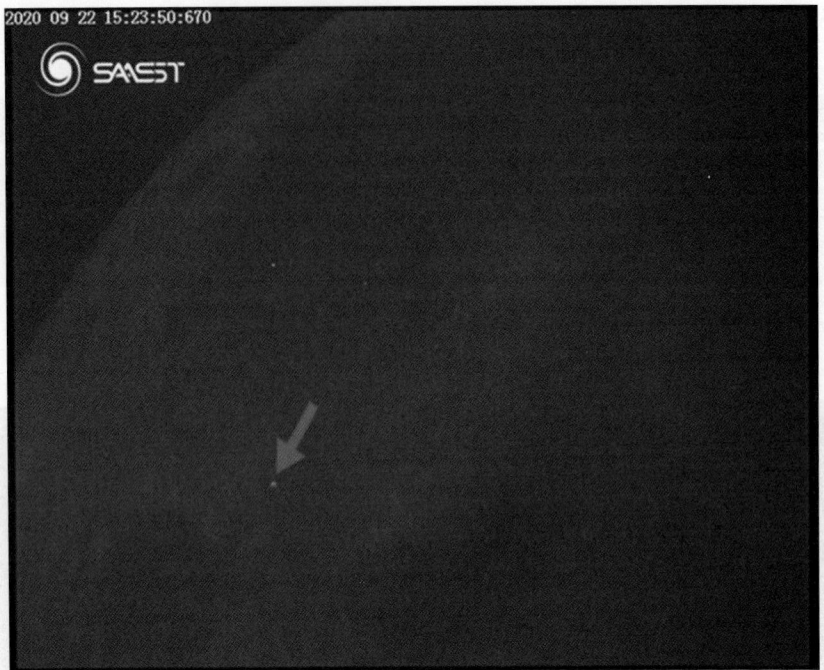

Figure 11. Event 2: A captured frame by the Sharjah Lunar Impact CCD camera of the highest brightness during the lunar impact event on the 22nd of September 2020.

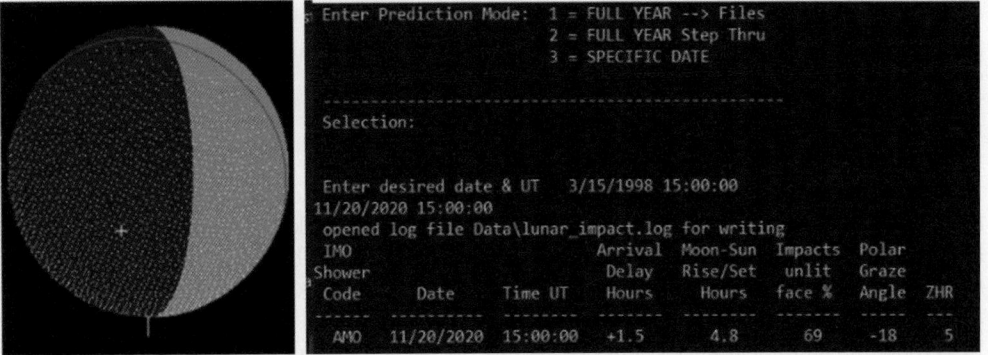

Figure 12. Event 3: The LunarScan 1.5 window of the lunar impact event on the 20th of November 2020 where it can be seen on the left where the software thinks the event took place on the Moon.

5.5. *Event 5*

This lunar impact event was detected on the 18th of December 2020 with a duration of 0.08 seconds, apparent magnitude of 7.79, mass of 0.14 kg, diameter of 0.08 meters and belongs to the Comae Berenicids meteor shower. Figure 13 shows the display window of the software for Event 5.

5.6. *Summary*

Table 1 displays a summary of properties for each event in addition to the lunar observation condition in which the detection took place in. From Table 1, it is clear that there is a considerable correlation between the illumination, elongation and age

Table 1. Properties of each lunar impact event and the associated Moon conditions.

		Event 1	Event 2	Event 3	Event 4	Event 5
	Date	01-03-20	22-09-20	20-11-20	20-11-20	18-12-20
Event Properties	Meteor Shower	Delta Leonid Complex	Southern Delta Piscids	Alpha Monocerotids	Alpha Monocerotids	Comae Berenicids
	Duration (s)	–	0.08	0.08	0.04	0.08
	Apparent Magnitude	7.94	7.94	9.54	10.06	7.79
	Mass (kg)	1.92	0.24	0.06	0.017	0.14
	Diameter (m)	2.84	0.09	0.06	0.04	0.08
Moon Conditions	Illumination	–	34%	33.80%	33.80%	33.80%
	Elongation	–	71°	70.57°	71°	71°
	Age (Days)	–	5.9	5.9	5.9	5.9

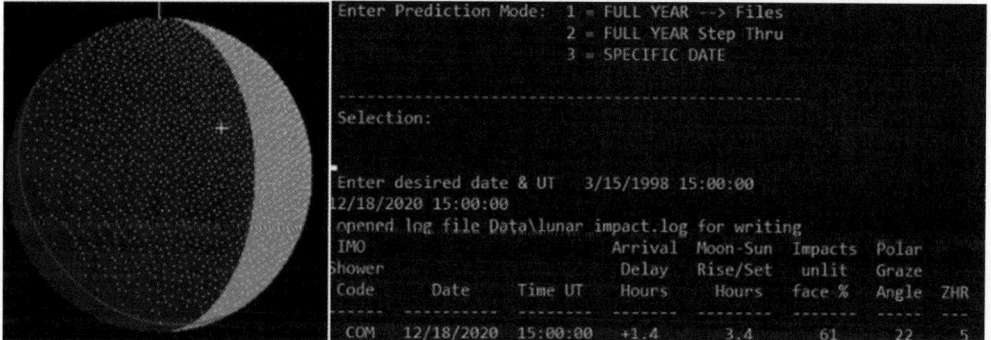

Figure 13. Event 5: The LunarScan 1.5 window of the lunar impact event on the 18th of December 2020 where it can be seen on the left where the software thinks the event took place on the Moon.

of the Moon with the probability of detections, as they are almost unified. As these events are during the first year of operation for the observatory, it is fair to assume that these lunar conditions are as close as "ideal" as they can be for detection of lunar impact events given the instrument capabilities and observational skills as the detection process takes some getting used to. Also, it is very important to note that there were different educated guesses/assumptions of distance, density and velocity for each event which then yielded information about the impactors. The impact flash magnitudes were calculated via determining the apparent magnitude of a nearby star in the same frame and comparing it to the flux of the impact using a light measurement tool created by Limovie in which the photometric signal of both the star and the impact are compared.

6. Conclusion

The Sharjah Lunar Impact Observatory (SLIO) has been operating from the campus of Sharjah Academy of Astronomy, Space Sciences and Technology (SAASST) since the year 2020. The coordinates of the site are 25°17'02.1"N 55°27'48.4"E. The observatory employs a 14" LX 200 ACF Schmidt-Cassegrain Meade telescope, with a focal ratio of 10 (f/10). The camera utilizes a 0.5" monochrome CCD Sony ICX439ALL sensor and is a high-sensitivity Watec video camera of the model 902H Ultimate recording at 25fps with a resolution of 720 x 576 pixels. In this paper, we presented details of the the observatory's instruments, observation and analysis processes and details of 5 events detected by SLIO in the year 2020. The analysis concluded that the impacts detected had apparent magnitudes of 7.94, 8.92, 9.54, 10.06, and 7.79, associated durations of

0.04, 0.08, 0.08, 0.04, 0.08 seconds and masses of 1.92, 0.24, 0.06, 0.017, 0.014 kilograms respectively.

References

Ait Moulay Larbi, M., Daassou, A., Bouley, S., Baratoux, D., Benkhaldoun, Z., & Lazrek, M. 2013, *First lunar flashes detected from Morocco at AGM observatory of Marrakech.* https://meetingorganizer.copernicus.org/EPSC2013/EPSC2013-333-2.pdf

Avdellidou, C. 2018, *Impact flashes on the Moon* https://meetingorganizer.copernicus.org/EPSC2018/EPSC2018-299.pdf

Baratoux, D., Colas, F., Chennaoui Aoudjehane, H., & Zouhair, B. 2012, *A French-Morrocon Project for the Studies of Impact Processes on the Earth and the Moon.* https://www.researchgate.net/publication/230667013_A_French-Morrocon_Project_for_the_Studies_of_Impact_Processes_on_the_Earth_and_the_Moon

Cook, T. (n.d.). Lunar Impact Flash Toolkit & Upload Interface. In *Grande, M. Prifysgol Aberystwyth University.* Retrieved December 28, 2021, from http://planetaryspaceweather-europlanet.irap.omp.eu/pdf/9_Cook_Impact.pdf

Madiedo, J., Ortiz, J., Organero, F., Ana-Hernández, L., Fonseca, F., Morales, N., & Cabrera-Caño, J. 2015, Analysis of Moon impact flashes detected during the 2012 and 2013 Perseids. *Astronomy & Astrophysics.*

Meade. (n.d.). 8", 10", 12", 14", 16" LX200®-ACF 2021, *Advanced Coma-Free Telescopes with GPS and AutoStar® II Hand Controller: Instruction manual.* Retrieved December 28, 2021, from https://www.bhphotovideo.com/lit_files/514162.pdf

Mohon, L. 2000, *About Lunar Impact Monitoring.* NASA. https://www.nasa.gov/centers/marshall/news/lunar/overview.html

NELIOTA. (n.d.). 2022, *Lunar Impact Events (NELIOTA - Home).* Retrieved January 27, 2022, from https://neliota.astro.noa.gr

Nittler, L.R., Alexander, C.M.O'D., Gao, X., Walker, R.M., & Zinner, E.,2021 *WAT-902H2 ULTIMATE/WAT-902H3 ULTIMATE: Operation manual.* Retrieved December 29, 2021, from http://www.altavision.com.br/Arquivos/Watec/Manuals/WAT-902H3-Ultimate_Manual.pdf

Talafha. M. (n.d.). Sharjah Lunar Impact Observatory (SLIO). In *Sharjah Academy for Astronomy, Space Sciences and Technology.*

Talafha, M., Abdulla, A. E. M., Azmi, M. M., Al-Naimiy, H. M., Al-Wardat, M. A. (2022), Sharjah Lunar Impact Observatory (SLIO). https://iopscience.iop.org/article/10.1088/1748-0221/17/04/T04008/meta

Talafha, M., Fernini, I., al Naimiy, H., Chogle, F., & Rajawat, A. 2020, *Lunar-Flashes.* https://iafastro.directory/iac/archive/browse/IAC-20/A3/VP/57766/

Astronomical Hazards for Life on Earth
Proceedings IAU Symposium No. 374, 2025
G. Tancredi, ed.
doi:10.1017/S1743921324000711

The meteoroid component of the astronomical hazard to life on Earth: contribution, relationships and more

Svitlana Kolomiyets⬭, **Mykola Kaliuzhnyi**, **Valeryi Zarytskyi**
and Iryna Kyrychenko

Kharkiv National University of Radio Electronics,
Nauky ave.14, Kharkiv, 61166 Ukraine
email: `svitana.kolomiyets@nure.ua`

Abstract. The ecological well-being of the Earth is closely connected with the prevention of asteroid-comet-meteoroid hazard. Asteroids and comets are the parent bodies of many meteoroids. Meteoroids, which are observed as meteors in the Earth's atmosphere, always collide with the Earth. This means that the orbits of such meteoroids can be a signal for the detection of potentially dangerous larger bodies in such orbits. At the same time, the dynamics of the complex of meteoroid orbits has a complex intricate character. Meteor science is also engaged in unraveling the patterns of orbital paths and paths of interplanetary bodies potentially dangerous for life on Earth. A separate section of meteor science is associated with the chemistry of the influx of meteoric matter. The report is devoted to the analysis of the above problems, as well as related issues, using open databases of meteor data and others, with an emphasis on radio data.

Keywords. Meteors, meteoroids, astronomical hazard to life on Earth, meteor radio observations

---

1. Introduction

Asteroid-comet-meteoroid hazard is of the types from the list of astronomical danger to life on Earth. Asteroids, comets and meteoroids are objects of the Solar system that are classified as Small Bodies. Meteoroid is a solid natural object of a size roughly between 30 μm and 1 meter, moving in, or coming from, interplanetary space (according 2006 IAU definition). In the IAU 1961 terminology, a meteoroid is an object smaller than an asteroid (or comet), but larger than an atom (or molecule). In addition, a naturally occurring body, even larger by modern IAU standards than a meteoroid, that moves through the Earth's atmosphere, having arrived from space, is considered a meteoroid. If this body (meteoroid) was large enough so that it did not burn up in the Earth atmosphere, but reached the Earth's surface, then the fallen body is called a meteorite. When large objects impact terrestrial planets, there can be significant physical and biospheric consequences. Impact events on the Earth are possibly related to the following events:
- formation of the Earth–Moon system;
- building blocks for life (panspermia);
- origin of water on Earth;
- several mass extinctions (1st: the prehistoric Chicxulub impact, 66 million years ago; 2nd: Cretaceous–Paleogene extinction event but acceleration of the evolution of mammals).

They undeniably caused such well-known catastrophic events as the Tunguska event, which occurred in Siberia, in 1908 and the 2013 Chelyabinsk meteor event.

How is the danger of invasion of Earth atmosphere and impacts on Earth's surface from space bodies estimated?

There are theoretical estimates of the frequency of possible impacts on the Earth of cosmic bodies, depending on their size (Kruchynenko (2012)):

- Stony asteroids with a diameter of 4 meters enter about once a year;
- Asteroids with a diameter of 7 meters enter about every 5 years;
- Asteroids with a diameter of 20 m strike Earth twice every century;
- 500 meteorites reach the surface each year; 5-6 tons per day; 2,000 tons per year;
- Objects with a diameter less than 1 m (meteoroids) : > 50,000 tons per year.

There are other estimates for meteoroids, e.g. (Hughes, (1992)), (Tancredi, (2019)). According (Tancredi, (2019)) every year, the Earth is hit by about 6,100 meteoroids large enough to reach the ground as meteorites, or about 17 every day.

There are two known risk assessment scales - the Turino scale and the Palermo scale. The Palermo scale is more complex and is used for meaningful scientific assessments of collision risks. About 19 Potentially Hazardous Asteroids (PHAs) are listed on the Sentry Risk Table (August 2021), but only none of them is actually critical. Asteroid Apophis removed from the list of extremely dangerous after clarification of its orbit. Only one asteroid is currently estimated to put the Earth at high risk of impact, and that is Bennu. As of June 2022, there are 2,270 known PHAs. This is about 8% of the total population around Earth - NEAs (Near Earth Asteroids).

The protection of the Earth from cosmic impacts began in the 20th century with the establishment of regular observations of the sky for the purpose of cataloging and determining the parameters of already discovered and undiscovered new asteroids and comets. We draw attention to a new step in approaches to the protection of the Earth in the 21st century, namely the transition from observations to concrete action. This is NASA's DART mission (Asteroid Deflection Mission), which was conducted already in September, 2022 (Trigo-Rodríguez, (2022)).

Which points on Earth are most prone to the danger of a meteorite fall? There is no answer to this question yet. There is the beginning of an effort to assess the impact risk on the Earth using backward-integration methods (Zuluaga & Sucerquia (2018)). This new method for calculating the relative probability of an asteroid or meteoroid colliding with the surface of a Solar system body is proposed.The largest probabilities predicted by this method "are localized on extensive geographical areas around to geographic apex equator (locations located in directions perpendicular to the apex-antapex direction). Chelyabinsk and Tunguska areas are located during most part of the year at low apex latitudes (large geographic apex colatitudes), where impact fluxes are larger".

There are "experimental" results of impacts on the Earth, these are the so-called wounds of the Earth - astroblems and found meteorites (McSween, (1999)). At least 200 astroblems are known. Among them, the oldest and largest is Vredefort, SA (the age $\sim$ 23 mya, D $\sim$ 190 km). Associated with the extinction of dinosaurs is Chicxulub, Yucatan, Mexico (the age   65 mya, D $\sim$ 150 km). The "best-preserved meteorite crater on Earth" is the Arizona Crater in the USA. The crater was formed approximately 50 000 years ago during the Pleistocene epoch, its D $\sim$ 1.2 km. Hoba(Namibia) is a tabular body of metal, measuring $2.7 \times 2.7 \times 0.9$ m. This meteorite is composed of about 84 % iron and 16 % nickel, with traces of cobalt.The main mass is estimated at more than 60 tons. The Murchison meteorite (Australia) is the oldest stone on Earth. It may be older than the Earth and the Solar system by 2.5 billion years. Australian meteorite one of three with key building blocks for life. This could be important evidence that life on Earth could have been introduced from space (Glavin, Elsila & et al. (2021)). This meteorite is well known and valuable in that its flight and fall were observed and recorded.

As of July 14th, 2022, 41 meteorites with photographic orbits were known for which orbital parameters were published. According to the date of fall, the Přibram meteorite is the first in this list. This list of meteorite orbits began its formation with the deployment of meteor observation networks (Borovička & et al. (2003)).

There are 7 Astroblems and 45 meteorites found on the territory of Ukraine. Seven recognized astroblems of Ukraine:. Obolon', Rotmistriv, Boltysh, Zeleny Gai, Ternovka, Bilylska, Illyinets (Kelley & Gurov (2010), Vidmachenko, (2017), Vidmachenko, (2018)). Among them it should be noted Boltysh (D $\sim$ 24 km, Kirovohrad Oblast, Ukraine, $48°45'N\ 32°10'E$) and Obolon' (D $\sim$ 20 km, Poltava Oblast, Ukraine, $49°35'N\ 32°55'E$). These two craters are among the largest (with a diameter of 20 km or more) confirmed craters listed in the 2017 Earth Impact Database.

Among the 45 meteorite falls and finds known on the territory of Ukraine, four belong to meteorite showers: Knyahinya, Zhovtnevyi Khutir, Krymka, Chervyn Kut. 'Knyahinya' (stone; ordinary chondrite L/LL5) is the largest meteorite found in Europe in recent history. Coordinates: $48°54'N$, $22°24'N$. Meteorite fall (1866) with a total mass of more than 500 kg. It is also known for the presence in its composition of possible traces of primitive extraterrestrial life. The main mass of is stored in a museum in Vienna, Austria.

According to the definition, a cosmic body 2 mm in size that fell on the Earth's surface is still considered a meteorite. In addition to meteorites, micrometeorites fall to the Earth's surface. A micrometeorite is a micrometeoroid that has survived entry through the Earth atmosphere with a diameter of 12-700 μ (uMM - unmelted micrometeorites, CS - cosmic spheres). Such estimates are known, the pre atmospheric flux: 15,000 tons/year, in atmosphere uMM and CS : 1,600 and 3,600 tons/year, respectively. For over a century, scientists only looked for micrometeorites in exceptionally clean and remote areas, such as the ice of Antarctica. The Earth Observation Data confirmed that over 100 metric tonnes of micrometeorites enter into the Earth's atmosphere daily, but searching for stardust in an urban area was still considered to be a futile task. A small subset of scientists collected enough micrometeorites from the Antarctic to create a substantial reference catalog. Their work set the stage for Jon Larsen's groundbreaking discovery of the world's first urban micrometeorite (NMM 1 - a barred olivine spherule called Brevik) in Norway in 2015 (Project Stardust, (2022), (Project Stardust (2022)), https://projectstardust.xyz/2022/06/17/stardust-could-have-provided-the-building-blocks-of-life/).

The search continues for evidence that the building blocks needed to build protocells in the primordial soup of ancient Earth are common in space. More than 70 different complex organic molecules have been found in space rocks. These are carbon-rich meteorites (known as C-chondrites) found on Earth. For example, the Murchison meteorite and the first micrometeorites discovered in Antarctica. Similar building blocks have also been found in space dust brought back to Earth by Japan's Hayabusa 2 probe (https://en.wikipedia.org/wiki/Hayabusa2). The question of the presence of these building blocks of life in sufficient quantities of micrometeorites and the remains of meteoroids in their precipitating traces after the combustion of a meteoroid in the Earth atmosphere remains open.

The possibility of easy introduction of life to planets with the arrival of meteoric matter or micrometeorites, if this hypothesis is confirmed, would at the same time imply a similar possibility of introduction of unwanted or life-threatening biological content from space to the planet.

Astroblems and meteorites on Earth, events such as Tunguska and Chelyabinsk were the results of collisions with the Earth of an asteroid or comet, sometimes, perhaps, a

Table 1. The selectivity of detection methods to meteoroid mass (Cheplecha (1998))

Detection Method	Mass Range	Magnitude Limit
photographic (small camera)	$10^{-1}...10^{2}$	+2
photographic (Super-Schmidt)	$5*10^{-4}...10^{0}$	+4
TV	$2*10^{-5}...10^{-3}$	+10
radar	$10^{-7}...10^{-3}$	+8...+14

lunar or earth-like body (during the formation of the Solar system and the Earth-Moon system). In some cases, the classification is obvious as to whether a comet or an asteroid was the cause of one or another impact event, and in some cases this is a controversial issue. Similar difficulties apply to more modest cosmic intrusions into the Earth atmosphere in the form of meteoroids of standard size in IAU 2006 terminology, only then we are talking about a comet type or asteroid type. There are precise and well-defined indications for separating small bodies into comets or asteroids (or into cometary or asteroid types for standard size meteoroids), but at the same time there are numerous cases of difficulty or even impossibility of accurate identification, e.g. (Churyumov, & et al. (2010)). A significant part of meteoroids has genetic links with comets and asteroids as with parent bodies. Initially, such links were established between comets and meteor showers clearly associated with them. Subsequently, meteor showers with the parent body, an asteroid, were precisely established. Meteors and meteoroids that are not assigned to any meteor shower are called sporadic. Meteoroids, which are observed as meteors in the Earth's atmosphere, always collide with the Earth. This means that the orbits of such meteoroids can be a signal for the detection of potentially dangerous larger bodies in such orbits. At the same time, the dynamics of the complex of meteoroid orbits has a complex intricate character. Meteor science is engaged in unraveling the patterns of orbital paths and paths of interplanetary bodies potentially dangerous for life on Earth (Gorbanev, Konovalova & Davruqov (2021)).

2. Data from the classical meteor specular radar "MARS" as a research tool

There is a special term "the meteor zone of the atmosphere" for the layers of the atmosphere on altitudes about 70-120 km, where meteors are observed mainly after the entry of meteoroids. The integral influx of meteoric matter on the Earth can be represented as such (Kruchynenko (2012), Kolomiyets *et al.* (2016)).

$$lgN_R = -7.86 - 0.892lgm.$$

where N_R - the influx of bodies with mass m per year for the entire Earth in the mass range of $10^{-18} - 10^{22}g$. The daily top estimate of the number of meteoroids with masses under 0.01 g, burning in the upper atmosphere over the whole Earth, is given 200 million events. The selectivity of detection methods to meteoroid mass is presented (Table 1). Experimental meteor orbital data are provided by classical ground-based remote methods for observing meteors in the atmosphere Earth: visual, optical, photographic, TV and radar, as well as in situ.

MARS - Kharkiv meteor radar system (Ukraine) has acquired the status of the important historical astronomical instrument (Table 1) in world history (Fedynsky *et al.* (1976), Kolomiyets *et al.* (2012)). It still remains one of the most highly sensitive specular reflection meteor orbital radar systems. The system could register meteors up to + 12 magnitude, the range of recorded masses was from 10^{-6} to 10^{-3} of a gram.

Meteors are normally arriving at random, which means that their number in a given interval follows the Poisson distribution. Hence, any measures based on meteor counts (be it the counts themselves, or, e.g., total reflection duration for that number of counts) will exhibit a spread from the "true" value. In addition, the received meteor data do not fully and selectively reflect the true picture. When obtaining estimates of the distributions of meteoroid orbital elements, it is necessary to take into account the selectivity factors of radar measurements: physical, geometric, astronomical, and instrumental. When processing the results of the MARS meteor radar, the following mathematical expressions were used to take into account the selectivity factors (Kashcheyev & Tkachuk (1980)).

Selectivity factors:

Geometric factor: $W_1 = \int_0^{24h} S_{ef}(\delta, t)dt$

Physical factor: $W_2 = \frac{1}{l_0} \int_0^{\infty} l(M_0, V_0, Z, \alpha_{ef}^{min})dM$

Astronomical factor: $P_3 = \frac{\pi * V_g * sin(i)}{R_{SC}^2 * V_g^2} * (2 - \frac{1}{a} - p)^{\frac{1}{2}}$

There are many other selectivity estimates and all of them can be a target for criticism.

The MARS meteor radar had two subsystems: a subsystem for recording the number of meteor events and an orbit recording system (Kashcheyev et al. (1977)). When registering the number of meteors, the meteor events could reach 1000-5000 meteors per hour. With such a flow of events in the MARS meteor radar, the collection and processing of radio meteor information was carried out to determine the meteoroid density flux and solve a number of other problems. Note that the data from the MARS meteor radar were used to build an engineering model of the meteoroid environment.

In the 21st century, new opportunities have emerged associated with the development of radio electronics and the information technology industry. All-sky meteor systems "SKiMET meteor systems" (Hocking, Fuller & Vandepeer (2001)) have become widespread, and not so long ago another modification of them, ComMet/21i (abbreviated as "ComMet"), appeared. Two meteor orbital radars based on SKiYMET technology operate in Canada - "CMOR" (Brown & et al. (2004)) and Argentina - "SAAMER" (Janches, Hormaechea & Brunini (2021)). They register meteoroid orbits, the survey of which is used to build a model of the meteoroid environment of the US National Aerospace Agency - NASA.

A comparison was made of the distributions of the main parameters of the orbits and velocities of the Ukrainian meteor radar "MARS" and other meteor radars from the (Table 2). The distributions in the main regularities are consistent. This can be seen as an additional argument to confirm the satisfactory quality of the MARS meteor radar data for what was planned in the current study. The velocities of meteorites that have fallen on Earth are usually less than 23 km/s. In (Kolomiyets *et al.* (2020)) it was shown some features of the distributions of meteor radio data samples for geocentric velocity Vg less than 23 km/s. This study was continued and supplemented.

3. Results & Problems

The orbital dynamics of celestial bodies (in particular, meteoroids) is to a certain extent determined by astrochemistry. According to radar observations of meteors, it is not possible to directly determine the chemical composition and density of registered individual meteoroids. Therefore, indirect methods are used, assumptions and models are developed, and criteria are introduced when researching data on the orbits of radio meteors. In particular, in the conducted research, the division of orbits into comet and asteroid orbits was made according to the Tisserand criterion, the heights of meteor trails in the Earth's atmosphere were estimated, the configuration was analyzed and the

Table 2. Comparison MARS with others well-known classic meteor radars: XX century AMOR (Baggaley *et al.* (1994) and HRMP (Hawkins (1963)); XXI century CMOR and SAAMER

Country	Radar name	Radar type	Method	Frequency	Site	LAT	LON	Observation	Orbits	Magnitude
Ukraine	MARS	VHF	mirror reflect Fresnel oscillation	22.38 MHz	Kharkiv	$49°24'50''N$	$36°52'E$	1967-1971	90 000	$+12^m$
Ukraine	MARS	VHF	mirror reflect Fresnel oscillation	31.1 MHz	Kharkiv	$49°24'50''N$	$36°52'E$	1972-1978	250 000	$+12^m$
New Zeland	AMOR	VHF	mirror reflect hybrid approach to interferometry	26.2 MHz	Bank Peninsula	$43°6S$	$172°.6E$	1995-1999	500 000	$+13^m$
USA	HRMP	VHF	mirror reflect Fresnel oscillations	41.1 MHz	Havana, Illinois			1959-1971	10 000	$+12^m$
Canada	CMOR	VHF SKiYMET	SKiYMET interfometry	29.85 MHz	Tavistock,	$43°.3N$	$-80°.8E$	May 2002 till now	> 5 000 000	$+8^m$
Argentina	SAAMER	VHF SkiYMet	SKiYMET interfometry	32.55 MHz	Tierra del Fuego	$53.68S$	$67.87W$	2010 till now	> 1 000 000	$+8^m$

features of the space of meteor orbit parameters were identified for selected data samples, and comparisons were made. To divide meteoroid orbits into comet and asteroid orbits, we used the Tisseran constant according to (Kresak (1969))

$$T_j = a^{-1} + 0.1686a^{l/2}(l - e^2)^{l/2}cosi_0, \tag{1}$$

where i_0 is the inclination of the meteoroid orbit with respect to Jupiter orbital plane, a is the semi-major axis and e is eccentricity of the meteoroid orbit. The boundary between stable asteroidal orbits and unstable cometary orbits at $i_0 = 0^0$ corresponds to $T_j = 0.58$. The typical value for asteroids is $T_j > 0.58$ and for comets is $T_j < 0.58$.

In the MARS meteor radar database (6460 orbits in 1978), the division of orbits into asteroid and comet orbits according to the Tisserand criterion showed the following statistics: 73% of the asteroid and 27 comet types, for speeds less than 23 km/s 69.7% and 30.3%, respectively.

Figure 1 presents some of the results of the study. Diagrams a-e are informative from the point of view of studying possible collisions of cosmic bodies with the Earth. Figure 1b shows the lines limiting the possibility of observing meteors on Earth in the space of orbital parameters. They are: $a = 1/(1 - e)$; $a = 1/(1 + e)$. A comparison was made for the e-a diagrams for objects whose orbits intersect with the Earth's orbit: Figure 1a) Chelyabinsk meteoroid (ChM) and AAA asteroids, Credit : Villatoro, (2013); Figure 1b) almost all currently known NEAs (in 2014) are shown by dots here. Circles represent NEC and some sh-P comets; Figure 1c) sporadic meteors of MARS radar data with inclinations from 0 to 30 degrees with a histogram of their eccentricities. Credit: Voloshchuk & Kashcheev (1996); Figure 1d) sporadic meteors (MARS radar data in 1978), asteroid type; Figure 1e) sporadic meteors (MARS radar data in 1978), comet type, Vg < 23 km/s (aphelion limited: semi-major axis up to 24 AU); Figure 1f) sporadic meteors (MARS radar data in 1978), asteroid type, Vg < 23 km/s; Figure 1g) sporadic meteors (MARS radar data in 1978), comet type, Vg < 23 km/s, aphelion Q limited to 20 AU, semi-major axis up to 2.4. Height distributions of reflecting points of meteor trains for meteoroids of asteroid (Figure 1h) and comet (Figure 1i) types with Vg < 23 km/s with aphelia 1.5 < Q < 3.8 AU (MARS radar data in 1978) were compared. A difference was found. It was confirmed that for a sample of sporadic meteors with velocities less than 23 km/s, all orbital inclinations are less than 60 degrees.

4. Conclusions

• It was shown that meteors, meteoroid orbits, chemical and other properties of meteoroids are an important independent and integral part of both the astronomical sky and the astronomical well-being and danger to life on Earth.

• Near-Earth meteoroid orbits (their statistics and distribution features) are an important tool for studying various regions of the Solar system and the movement of meteorites.

• In the conducted research, the division of orbits into cometary and asteroid orbits was made according to the Tisserand criterion, the heights of meteor tracks in the Earth atmosphere were estimated, the configuration and spatial features of the parameters of meteor orbits were identified for selected data samples, and comparisons were made.

• The following statistics has been received: 73% of the asteroid and 27 comet types, for speeds less than 23 km/s 69.7% and 30.3%, respectively. Orbits with velocities Vg < 23 km/s turned out to be 11%. It was confirmed that for a sample of sporadic meteors with velocities less than 23 km/s, all orbital inclinations are less than 60 degrees.

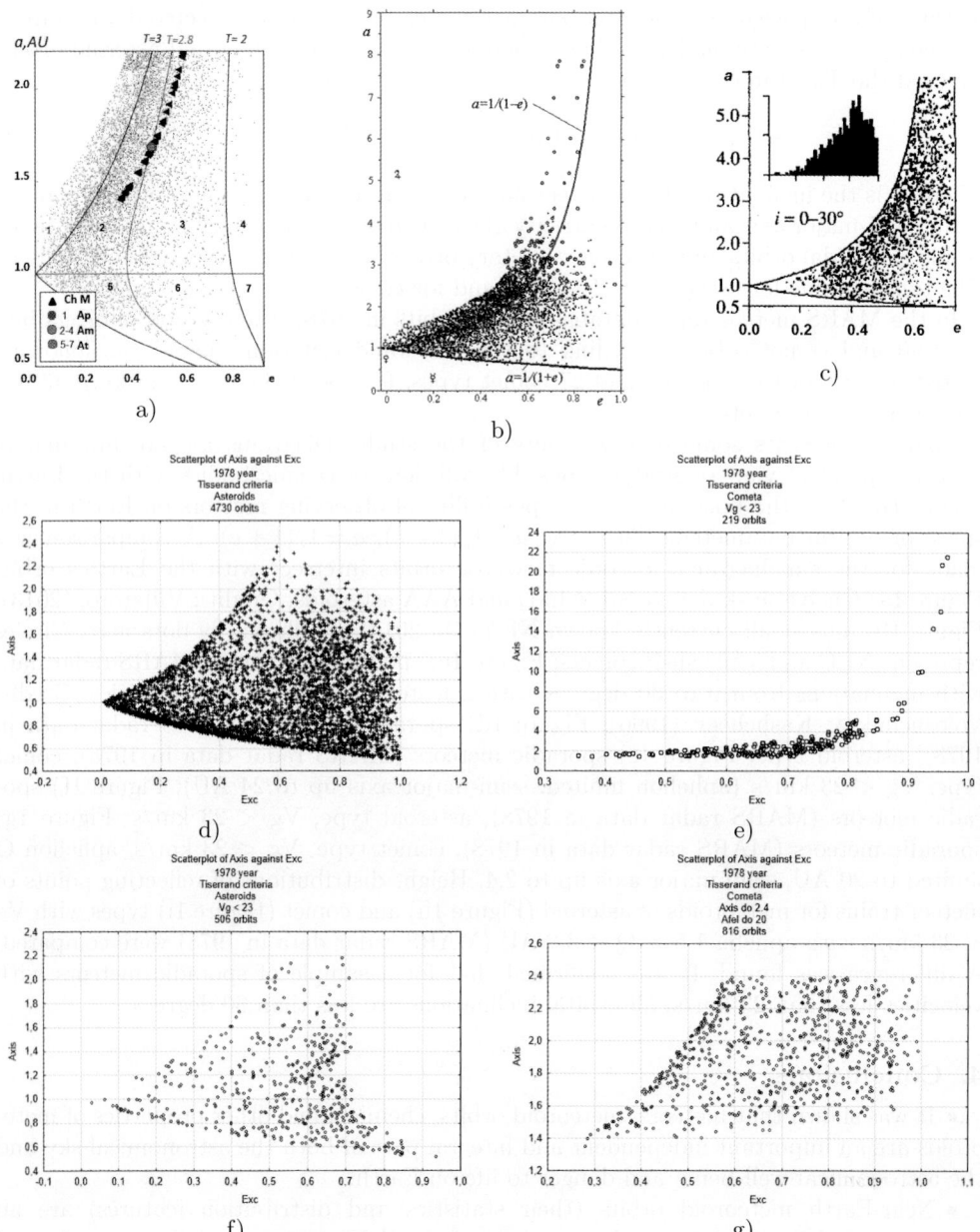

Figure 1. The e-a diagrams for objects whose orbits intersect with the Earth's orbit: a) Chelyabinsk meteoroid (ChM) and AAA asteroids, Credit: Villatoro, (2013); b) almost all currently known NEAs (in 2014) are shown by dots here. Circles represent NEC and some sh-P comets; c) sporadic meteors of MARS radar data with inclinations from 0 to 30 degrees with a histogram of their eccentricities. Credit: Voloshchuk & Kashcheev (1996); d) sporadic meteors (MARS radar data in 1978), asteroid type; e) sporadic meteors (MARS radar data in 1978), comet type, Vg < 23 km/s (aphelion limited: semi-major axis up to 24 AU); f) sporadic meteors (MARS radar data in 1978), asteroid type, Vg < 23 km/s; g)sporadic meteors (MARS radar data in 1978), comet type, Vg < 23 km/s, aphelion Q limited to 20 AU, semi-major axis up to 2.4. Height distributions of reflecting points of meteor trains for meteoroids of asteroid (h) and comet (i) types with Vg < 23 km/s with aphelia 1.5 < Q < 3.8 AU (MARS radar data in 1978).

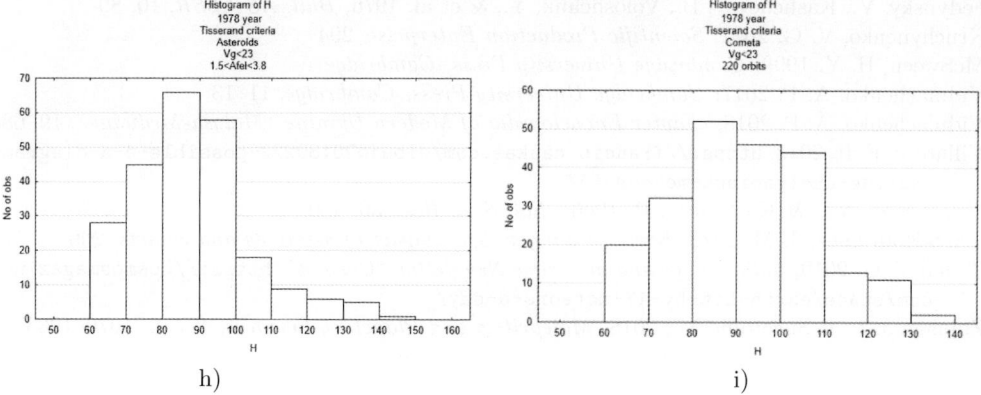

h)   i)

Figure 1. Continued.

• MARS still remains one the most highly sensitive specular reflection meteor orbital radar systems with one of the biggest meteor databases in the world. It is planned to modernize MARS using SKiYMET technology.

• Meteoroids, which are observed as meteors in the Earth's atmosphere, always collide with the Earth. This means that the orbits of such meteoroids can be a signal for the detection of potentially dangerous larger bodies in such orbits.

5. Acknowledgments

The work was carried out within the framework of state budget research projects (0121U109792) and dissertation research of the Kharkiv National University of Radio Electronics of the Ministry of Education and Science of Ukraine.

References

Baggaley, J., Bennett R., Steel D., & Taylor A. 1994, *Quarterly Journal of the Royal Astronomical Society*, 35, 293

Borovička J., Spurný P., Kalenda P. & Tagliaferri, E. 2003, *Meteoritics & Planetary Science*, 38, 975–987

Brown, P., Jones, J., Weryk, R. J., & Campbell-Brown, M. D. 2004, *Earth Moon Planet*, 95, 1, 617

Cheplecha, Zd. & et al. 1998, *Space Science Reviews*, 84, 327–421

Churyumov, K., Kruchynenko, V., Chubko, L., & Churyumova, T. K. 2010, *The International Astronomical Union*, 263

Glavin, D. P., Elsila, J. E., McLain, H. L. & et al. 2021, *Meteoritics & Planetary Science*, 56(1), 148–173

Gorbanev, Yu. M., Konovalova, N. A., & Davruqov, N. Kh. 2021, *Journal of physical studies*, 25, 4, 6

Janches, D., Hormaechea, J. L., & Brunini, C. 2013, *Icarus*, 223(2), 677–683

Hawkins, G. 1963, *Smithsonian Contributions to Astrophysics* 7, 53

Hughes, D. W. 2012, *Proceedings of the International Astronomical Symposium held at Smolenice*, 15

Kelley, S. P., & Gurov, E. P. 2002, *Meteoritics and Planetary Sciences*, 37, 1031–1044

Kolomiyets S. V. 2012, *CriMiCo 2012 - 2012 22nd International Crimean Conference Microwave and Telecommunication Technology, Conference Proceedings* , 94031, 46

Kolomiyets S. V. 2016, *Proceeding of IAU S 325*

Kolomiyets, S. V., Kolomiiets, K. A., Kyrychenko, I. Yu., Pryimachov, Yu. D., 2016, *Odessa Astronomical Publications*, 33, 109

Kresak, L. 1969, *Bull. Astr. Inst. Czechosl.*, 20, 4, 177–188

Fedynsky, V., Kashcheyev, B., Voloshchuk, Y., & et al. 1976, *Bull. AS USSR*, 10, 89

Kruchynenko, V. G. 2012, *Scientific Production Enterprise*, 294

McSween, H. Y. 1999, *Cambridge University Press, Cambridge*

Vidmachenko, A. P. 2017, *Cambridge University Press, Cambridge*, 11–13

Vidmachenko, A. P. 2018, *chapter Encyclopedia of Modern Ukraine. "Malysh-Medicine"*, 19, 688

Villatoro, F. R. 2013, https://francis.naukas.com/dibujo20130222-possible-e-a-diagram-for-the-chelyabinsk-meteoroid/

Voloshchuk, Yu., & Kashcheev, B. 1966, *Sol. Sysl. Res.*, 30, 480

Trigo-Rodríguez, J. M. 2022, *Earth in danger. The impact of asteroids and comets*, 200

Tancredi, G. 2019, *online magazine interview Newsletter "Cosmos"*, https://cosmosmagazine.com/space/earth-hit-by-17-meteors-a-day/

Zuluaga J. I., & Sucerquia M., 2018, *Meteoritics and Planetary Sciences*, 477, 2, 1970–1983

Astronomical Hazards for Life on Earth
Proceedings IAU Symposium No. 374, 2025
G. Tancredi, ed.
doi:10.1017/S1743921324000668

Jupiter and Evolution of Complex Life on Earth

Xuguang Leng ⓘ

freelance
Breinigsville, PA 18031, USA

email: xuguangleng@gmail.com

Abstract. Comets and asteroids collision with Earth and other planets is part of the continued planetary formation, the other part is the solar wind delivers water and gasses back to the Kuiper Belt from the planets, together they form the solar hydrologic cycle. The new theory of solar hydrologic cycle provides that solar wind stripping water and gasses from the inner planets, while having diminished effect on outer planets, is the cause for outer planets becoming gas giants. Jupiter, having the smallest orbit among outer planets, is destined to be the predominant planet and plays a critical role in complex life evolution on Earth. Jupiter grows mass by locking up comet mass, thus reducing the number of comet collisions to Earth. Reduced hydrogen infusion from comets enabled Earth's atmosphere transitioning from hydrogen to oxygen rich. The transition trajectories of Jupiter mass gain and Earth water and gasses mass fluctuation are calibrated using known geological events. Earth's trajectory can be divided into three periods, Hydrogen, Carbon Dioxide, and Oxygen, named after predominant gas in Earth's atmosphere for the period. Complex life only flourishes in the Oxygen Period, when aerobic metabolism is possible. Mass extinction can be caused by cometary hydrogen infusion that incinerates atmospheric oxygen. Probability of such astronomical hazards is declining as outer planets have locked up most comets and will continue to absorb more comets. Earth is safer than ever, and will become even safer, as dictated by the solar water cycle. The physics of the solar hydrologic cycle is universally true, life should be universal phenomena.

Keywords. solar wind, Jupiter, Earth, comet, evolution, planets formation.

---

1. Introduction

The end of humanity has been a topic of great concern across ages and civilizations. This is reflected in the wealth of references throughout many cultures and religions. Over the last several decades, studies have allowed us to better understand the most likely threats to life on Earth, both in the past and the future. Mass extinctions, which have thus far been caused by natural phenomena, have most substantially affected life on Earth.

What were the causes of previous mass extinctions, were they of solar activities, or were they of comet/asteroid collisions? We know with high confidence that one of them, the Cretaceous–Paleogene extinction event, was caused by the Chicxulub asteroid. What were the mechanisms? Were they radiation/magnetic? Collision impact? Or hydrogen infusion? Is the risk increasing or decreasing? We saw Shoemaker-Levy 9 collide with Jupiter in 1994, yesterday in astronomical terms. Is collision avoidance the only defense? Or are post-apocalyptic surviving measures feasible?

What was the condition in Earth's formative years? Was it hospitable to complex life? What was the process of transforming Earth's environment to being hospitable?

Going one step further from astronomical hazard, what about the evolution of life on Earth itself, was it influenced or even determined by astrophysics? Did ammonia and methane from comets sow the seeds for life on Earth, or these materials were recycled to comet from inner planets? Was Jupiter mass providing the shield for complex life to evolve on Earth? Then, what made Jupiter so massive in the first place? Chance? Or some physics were in play?

To further expand the horizon, can the mechanism/process that produced intelligent life on Earth produce life in other similar star-planet systems?

The new theory of solar hydrologic cycle attempts to address all these questions.

2. Solar Hydrologic Cycle

There is constant mass movement in the solar system. Mass is moving from Kuiper Belt to the inner solar system through comet collision with the planets, mass is also moving the opposite direction by solar wind stripping water vapor and gasses from the inner planets. The mass movement forms a cycle, like Earth water cycle, so it is called solar hydrologic cycle. The balance of the cycle determines the mass of the planets today, such that the solar hydrologic cycle is the continuation of planetary formation.

Hydrogen and other light elements are the most abundant elements in the solar system, rocky materials are a tiny fraction. Comparing total comet mass today of 2% solar mass Mendis et al. (1986), total asteroid mass of 12×10^{-10} solar mass Pitjeva et al. (2015) is negligible. The existence of a separate comet hydrogen envelope Mancuso (2015) indicates that the comet is hydrogen rich. Movement of mass in the solar system is the movement of lighter elements, primarily hydrogen.

Four billion years ago, in the young solar system, there were a vast number of comets, and outer planets were rocky dwarfs. The solar hydrologic cycle started with ferocious intensity. Comets rained on the planets, bringing large quantities of water and gasses to the planets.

Solar wind, on the other hand, works in the opposite direction. Solar wind brings hydrogen, water vapor and other gasses, back to the direction of Kuiper Belt, and at farthest, to the heliopause.

In Earth water cycle, water vapor stops at very low altitude to forms rain clouds even though water vapor is lighter than both nitrogen gas and oxygen gas. Solar system may have, yet to be discovered, a similar mechanism forming "comet clouds". In the region of Kuiper Belt to heliopause, the particles and molecules are bumping into each other forming small chunks, losing electric charge in the process. Ebb and wane of the solar wind, and gravity ripples of planets swinging by congregate the material like waves in the pond congregate leaves. The larger chunks are pulled inward by solar gravity to the Kuiper Belt, where they reform into comets by the positive feedback loop of gravity. There can be many cycles of water and gasses traveling between the Kuiper Belt and inner solar system over billions of years.

Solar wind exerts great influence in the inner solar system, blows away water and gasses from inner planets, including Mars (Barabas et al. 2007). This prevents inner planets mass growth from comet collisions. The solar wind, however, has diminished effect on the outer planets due to the distance. When comets collide with outer planets, their mass stays with the planets. Outer planets are deadends for solar hydrologic cycles. Using the Earth hydrologic cycle analogy, inner planets are like Amazon rainforest, outer planets are like Antarctic ice sheet. Amazon rainforest gets a lot of rain, but the water flushes back to the ocean. Antarctica is drier than the Sahara desert, yet holds 70% of Earth's fresh water.

Outer planets grow larger and larger as the solar hydrologic cycle repeats, locking up more and more comet mass, reducing the number of comets and intensity of solar

Table 1. Relationship of outer planets mass.

Outer Planet	Distance from the Sun [AU]	Mass [Earth mass]	Adjusted Orbital Mass Gain Constant	Predicted planet mass [Earth mass]
Jupiter	5.2	318	8599	
Saturn	9.6	95		93
Uranus	19.2	14.5		23
Neptune	30.0	17.1		10

hydrologic cycle, depriving inner planets of an ample supply of water and gasses. With solar wind constantly blowing, the inner planets began to dry up. Earth, with a magnetic field and large mass, is able to retain most of the water and heavier gasses.

The comet mass does not distribute to outer planets evenly or randomly, but in accordance with the probability of collisions. There are many factors that contribute to the probability. The comet orbit plane relates to the planet orbit plane is an important factor, yet does not favor any particular planet. Whether perihelion of comet orbit is inside the planet's orbit is another factor, yet minor for outer planets. The determining factor is the planet orbit size, which can be substituted with mean distance from the Sun. The planet mass gain is in inverse proportion with distance from the Sun. Assuming outer planets started with negligible mass, like one Earth mass, then, mass gain from the comet equals to planet mass. The equation can be expressed as:

$$m_1 r_1 - m_2 r_2 = c \tag{1}$$

Where m is the mass of a planet, r is the distance the planet is from the Sun. c is the Orbital Mass Gain Constant related to the total comet mass at Kuiper Belt.

The second weighty factor is the positive feedback loop where larger mass attracts more comet collisions, which in turn produces even larger mass. The positive feedback loop has many random factors, and can not be predicted by simple formulas. Since distance to the Sun drives mass, and mass drives positive feedback loop, the final mass could be driven by the square of the distance to the Sun.

Adding inverse relation with distance to the Sun as positive feedback adjuster, Eq. (1) becomes:

$$m_1 r_1^2 = m_2 r_2^2 = c_{adj} \tag{2}$$

Calculating the Adjusted Orbital Mass Gain Constant using Jupiter mass and orbit, then using the constant to predict the mass of other outer planets as in Table 1.

Saturn mass is perfectly aligned with Eq. (2) predicted mass. Uranus mass is smaller than predicted, perhaps because positive feedback loop effect is not as pronounced when the mass is small. Neptune mass is much larger than predicted, because of its proximity to the Kuiper Belt. There are many random factors influencing the planet's mass, so Eq. (2) is an approximation of outer planets mass relation.

Solar wind determines there can be no gas giant in the inner solar system, and mass gain from comet collisions determines outer planets mass is inverse with their orbit size. Both inner and outer planets mass is driven by the solar hydrologic cycle.

3. Jupiter Mass Growth Trajectory and Life Evolution on Earth

Now it has been established that ultimate planet mass is determined by solar hydrologic cycle, further investigation can be conducted into the trajectories of Jupiter mass growth and Earth water/gasses mass and composition. Then, the trajectories can be calibrated using the known geological events.

As Earth's cometary input is critically influenced by Jupiter's mass, Jupiter's mass growth trajectory needs to be determined first.

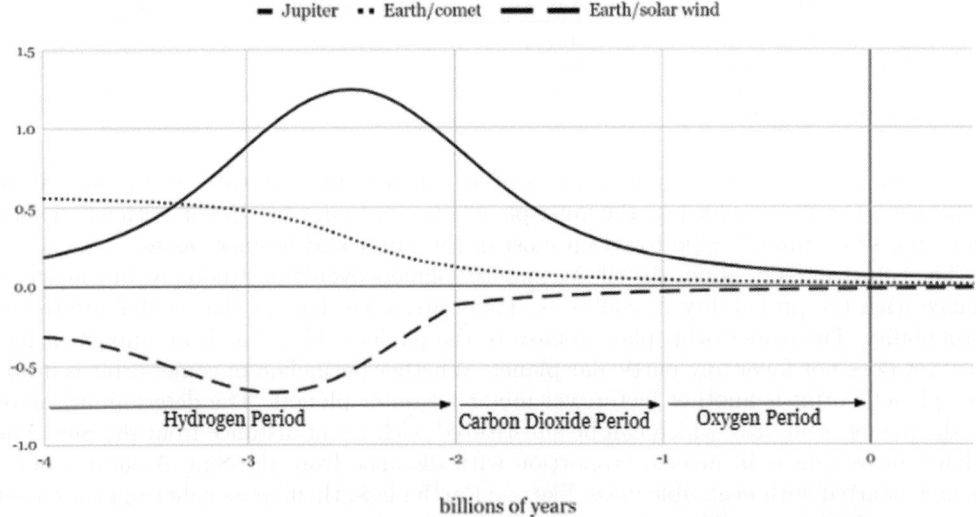

Figure 1. Jupiter mass gain, Earth mass gain and loss trajectories.

There is a finite number of comets, partially renewable. There is also a positive feedback loop, Jupiter mass growth time trajectory should follow the bell curve with three stages, early growth, peaking, and maturity. In early growth stage, there were a vast number of comets, yet Jupiter had a very small mass, probably Earth mass, most comet collisions happened in the inner planets. As Jupiter gradually gained mass, the positive feedback loop kicked in, Jupiter began to attract more comet collisions than its orbital share.

Jupiter entered the peaking stage when it absorbed more comets than the solar wind comet recycling had created. The total number of comets began to decline. On the one hand, Jupiter was taking more share of comet collision, on the other hand, the total number of comet collisions was declining. When the two lines crossed, Jupiter mass growth rate had reached the culmination point.

Jupiter entered the maturity stage when the growth rate dropped to the early growth stage level. See Figure 1.

Figure 1 is the theoretical prediction of regression curves of Jupiter mass gain. Comet collisions are discrete events both random on the time interval as well as size of the comet. A small percentage of large comets would have a disproportional influence on the curve assuming a normal distribution.

Jupiter mass growth determines Earth mass gain trajectory in two ways, firstly, Jupiter mass growth reduces total number of comet collisions, secondly, Jupiter takes more share of comet collisions. Earth mass gain rate decline is in sync with Jupiter mass growth, the trajectory is like an arc-cotangent curve, with sharpest decline occurring when Jupiter hit its growth culmination point.

Unlike Jupiter, Earth, as an inner planet, also has a minus side of the equation, solar wind stripping. The larger the air mass, the higher the lighter gasses ratio, the easier for solar wind to overcome Earth's gravity and magnetic fields, the higher the stripping rate. It takes time for mass from comet collision to accumulate, thus the solar wind minus trajectory should be mostly like a mirror image of plus trajectory with a lag. The strength of solar wind can also vary over time, it is a minor factor, however, thus not considered.

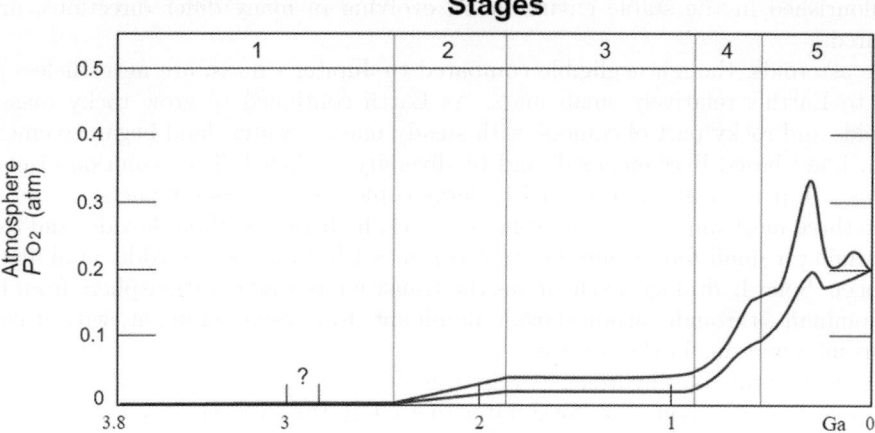

Figure 2. Estimated evolution of atmospheric PO_2 Holland (2006).

Corresponding to Jupiter's three stages, Earth has three periods, Hydrogen, Carbon Dioxide, and Oxygen, named after predominant gas in Earth's atmosphere for the period. Because solar wind blows away water and gasses, the three periods only differ on the intensity of comet bombardment, there is no dramatic change of Earth's mass.

The Hydrogen Period lasted approximately between 4 Ga (billion years ago) to 2 Ga. In the first half of this period, the vast numbers of comets, unimpeded by Jupiter's small mass, bombarded Earth and other inner planets, bringing in a vast amount of water and gasses. Earth was completely covered by water, and an air mass that could be several times of today's atmosphere. The air composition was hydrogen/helium majority, with some heavier nitrogen and carbon dioxide. Earth's meager gravity struggled to retain so much lighter gas, solar wind was blowing them away at a very fast rate.

The existence of a substantial amount of hydrogen deprives the atmosphere the opportunity to have more than a trace amount of oxygen. Any static electricity would ignite the hydrogen and consume the oxygen. In absence of oxygen, Hydrogen reacts with nitrogen and carbon dioxide to generate ammonia, methane, and other compounds, with the help of lightning. Earth was a gigantic "gas furnace".

In the second half of the Hydrogen Period, comets were increasingly intercepted by the growing mass of Jupiter. The gradual reduction of water/gas supply reached equilibrium with solar wind, Earth's air mass had peaked. There is, however, a large inventory of accumulated hydrogen to be blown away that could take a billion years. Heavier gasses like carbon dioxide and nitrogen were left behind and became a bigger and bigger part of the atmosphere. Oxygen still had to wait. See Figure 2.

Earth entered Carbon Dioxide Period at approximately 2 Ga, with most of the hydrogen blown away, leaving only heavy gasses like carbon dioxide and nitrogen in the atmosphere, the "gas furnace" had been shut down. The oxygen converted by photosynthesis began to accumulate, starting the Great Oxidation. Most of the oxygen was consumed by the oxidation of the Earth shell as well as by occasional comet hydrogen infusions, thus did not accumulate in large concentration in the atmosphere until approximately 1 Ga.

The Oxygen Period started about 1 Ga, when oxygen was unburdened by oxidation of the Earth shell and could accumulate in the atmosphere in significant concentration. Complex life could finally evolve as atmospheric oxygen makes aerobic metabolism possible. In this period, most loose objects had merged into the planets and Jupiter fully grown, the comet and asteroid collisions with Earth are few and far in between.

Life flourished in the stable environment, evolving in many differ directions, animals appeared.

The asteroids, though negligible compared to Jupiter's mass, are nevertheless significant to Earth's relatively small mass. As Earth continued to grow rocky mass from asteroids and rocky part of comets, with steady mass of water, land began to emerge on Earth. Land based lives emerged, and biodiversity exploded. The evolution of complex life was still periodically interrupted by large comet and asteroid impacts.

The three most important atmospheric gasses, hydrogen, carbon dioxide, and oxygen, are forming a small food chain. Oxygen is sourced from carbon dioxide, annihilated by hydrogen. Solar hydrologic cycle drives the transition of Earth's atmosphere from hydrogen dominant, through carbon dioxide dominant, to oxygen dominant, with occasional serious interruptions in the process.

The clock of the solar hydrologic cycle ticks slowly, so life evolution is a slow process. It took more than a billion years for Jupiter to grow to the sufficient size to be an effective "valve" to turn off unimpeded comet bombardment to the inner planets, thus allowing solar wind to blow away the excess hydrogen on Earth. It took another billion years for primitive life on Earth to produce enough oxygen to oxidize the Earth shell, thus allowing oxygen to accumulate in the atmosphere. Life has only been flourishing for the last billion years or so because it took more than three billion years to set the condition right, by solar hydrologic cycle.

The orbit location of primordial Jupiter influences Jupiter mass gain trajectory and ultimate mass. Should the orbit be farther out, Jupiter mass gain trajectory would flatten, culmination point delayed, and have a smaller ultimate mass. Earth's trajectory would decline slower and the Oxygen Period would arrive later, complex life would emerge later as well. Should the orbit be closer to the Sun, but without cross into inner planet orbit, Jupiter mass gain trajectory would steepen, culmination point advanced, and have an even larger ultimate mass. Complex life would emerge earlier on Earth. Should Jupiter cross into inner planet orbit, Saturn would have to take Jupiter's role as predominant planet. Saturn's larger orbit means it is less effective at absorbing comet mass, Earth would probably still be stuck in the Hydrogen Period today.

4. Astronomical Hazards

Comet collisions brought water and gasses to Earth, which enabled life. After life flourished, comet collisions became astronomical hazards, just like rain is the source of life, yet concentrated rainfall causes devastating flooding. It is inevitable that large, hydrogen rich comets have collided with Earth and caused anoxic events in the last billion years.

Coincidentally, there have been several major mass extinction events in the last billion years. The search for causes of extinctions has been focused on volcanism, and in combination with bolide impact Arens et al. (2008). The solar hydrologic cycle theory offers anoxic event by cometary hydrogen infusion as a more plausible cause for mass extinctions.

Compared to the vast abundance of hydrogen in the solar system, Earth's atmospheric oxygen mass is quite small. In today's atmosphere, the mass of oxygen gas is about 1.0×10^{18} kg. Since water by mass is 11% hydrogen, and 89% oxygen, it only takes about 1.2×10^{17} kg of hydrogen to consume all the oxygen in the atmosphere in a hydrogen-oxygen combustion. The Chicxulub asteroid mass is estimated at 1.0×10^{15} kg to 4.6×10^{17} kg Durand-Manterola et al. (2014). At the high estimate, if 25% of the mass is hydrogen, it can consume all the oxygen in today's atmosphere. Even if Earth's air mass in 0.5 Ga was several times of today's, and oxygen content doubled today's ratio of 21%, a hydrogen rich comet of size slightly larger than Chicxulub can still wipe out all the oxygen in Earth's atmosphere, causing anoxic mass extinction.

After Earth's oxygen mass was periodically fully or partially incinerated by large hydrogen infusions from comet collisions, it has to be rebuilt from carbon dioxide inventory or infusion from the same colliding comet. Photosynthesis is the predominant mechanism that produces atmospheric oxygen on Earth Dismukes *et al.* (2001), therefore maximum atmospheric oxygen mass is determined by the mass of carbon dioxide feedstock. Earth's oxygen level recovery does not always reach the previous peak.

Over time, Earth's carbon dioxide inventory is gradually scrubbed from the atmosphere by locking carbon into the Earth shell like forming coal, and releasing oxygen to the atmosphere. Later, derivative atmospheric oxygen is converted to water by hydrogen infusion. The air mass is becoming smaller, and the ratio of nitrogen is becoming higher.

Being essential to aerobic metabolism, the large, sudden fluctuations of the overall air pressure as determined by overall air mass, oxygen level, and carbon dioxide level, would offer advantage to some species while causing extinction to other species. Periodically, solar hydrologic cycle sets a new direction for life to evolve, homo sapien is a byproduct of one of the new directions. From humanity's point of view, the new direction was the right direction. In the grand scheme of things, an intelligent life could emerge from any of the directions.

On the positive side, Earth is currently in the period that the probability of comet collisions have greatly diminished. Jupiter has acquired a massive mass through absorbing most of the loose objects in the solar system. The intensity of the solar hydrologic cycle continues to weaken, Earth is safer than ever and continues to be even safer.

To fully mitigate the existential risk for humanity, comet/asteroid collision avoidance by diverting them through mechanical means is the best measure, though far from guaranteed success in today's technology. Modern day Noah's Ark, a fleet of nuclear powered submarines with long term oxygen generating capability scattered around the world, beached post-apocalyptic near river mouths and food storages, would be a superior doomsday "Plan B" measure. Submarines are low cost, high capacity, proven technology. They don't require ground support, which would be nonexistent post-apocalyptic. They are on the Earth, there is no landing risk. Space, moon, Mars based measures, if they are feasible at all, are far inferior.

5. Implications

Looking beyond Earth, did Mars host life in the past? Can Mars be a sanctuary for humanity? Like outer planets are mostly following Jupiter's mode, the inner planets are mostly following Earth's mode, with some minor differences. Mars, due to its smaller mass and larger orbit than Earth, gets less comet collisions. On the other hand, the solar wind is less intense due to farther distance from the Sun. On balance, Mars should be an abbreviated version of Earth. In the early years, Mars had a thick, hydrogen majority atmosphere, and liquid water. Mars' atmosphere could also have produced ammonia and methane. Mars' oxidized surface Hartman *et al.* (1995) indicates liquid water lasted at least to its carbon dioxide period that bacteria existed long enough to produce sufficient oxygen. Without the protection of big gravity and a magnetic field, had a smaller water/gas inventory and supply, Mars was blown dry by the solar wind not long after Jupiter reached peak growth rate. Liquid water returned to Mars intermittently when large comets collided, yet the dry spell is longer and longer. Mars can still have liquid water and a thicker atmosphere in the future, if only for a short period of time. Mars is no more hospitable than post-apocalyptic Earth.

Today's comets have a small amount of methane Mumma *et al.* (1996) and ammonia Wyckoff *et al.* (1991). Where do these volatile compounds come from? Methane, ammonia, and other volatile compounds were produced in inner planets during their respective hydrogen periods. The hydrogen period lasted a long time, so a large quantity of these compounds had been produced. Methane and ammonia are lighter than carbon dioxide and nitrogen, can be more easily stripped by solar wind and carried back to the Kuiper Belt, changing the composition of next generation comets. As comets containing these compounds fly into the inner solar system and collide with planets and their moons, the compounds are spread throughout the solar system.

The solar hydrologic cycle works like a clock, determining what time comet collision would tape off, what time Earth would have an oxygen rich atmosphere. The overall course of life evolution was pre-determined by physics, leaves no role for chance to play. Life on Earth was inevitable, complex life was inevitable, intelligent life was inevitable.

Extending horizon further beyond the solar system, does life exist in other parts of the universe? The physics of the solar water cycle is universally true. There are many solar system like planetary systems in the universe. They shall develop their own Jupiter-Earth pairs. Their earths shall also have water, oxygen rich air, land, long periods without catastrophic comet impact, thus evolution of complex life. Life should be universal phenomena, complex life should be universal phenomena, intelligent life should be universal phenomena.

References

Arens, N.C., West, I.D. 2008, *Paleobiology*, 34(4), 456-471

Barabash S., Fedorov A., Lundin R., Sauvaud J.-A. 2007, *Science*, 315, 501–503

Dismukes GC, Klimov VV, Baranov SV, Kozlov YN, DasGupta J, Tyryshkin A. 2001, *Proc Natl Acad Sci USA*, 98(5):2170-5

Durand-Manterola HJ, Cordero-Terceroar G. 2014, *astro-ph.EP*, arXiv:1403.6391

Hartman, H., McKay, CP. 1995, *Planet Space Sci*, 43(1-2):123-8

Holland HD. 2006, *Philos Trans R Soc Lond B Biol Sci.*, 361(1470): 903–915

Mancuso S. 2015, *A&A*, 578:L7

Mendis, D.A., Marconi, M.L. 1986, *Earth Moon Planet*, 36, 187–190

Mumma, M.J., et al 1996, *Science*, 272, 1310-1314

Pitjeva, E., Pitjev, N. 2015, *Proc. of the IAU*

Wyckoff, S., Tegler, S. C., Engel, L. 1991, *ApJ*, 368, 279W

Author index